THE ELFIN WORLD
OF MOSSES AND
LIVERWORTS

OF

MICHIGAN'S UPPER PENINSULA

AND ISLE ROYALE

Janice M. Glime, Ph. D.
Department of Biological Sciences
Michigan Technological University

Drawings by Marshall L. Strong

1993

Published by the
Isle Royale Natural History Association

TABLE OF CONTENTS

Dedicated to

Diana and Nathan
for teaching me that children
are not daunted by scientific names

ACKNOWLEDGMENTS

Although I have never taken a class from him, he has many times been my teacher. Without his help in identifications over the last 20 years, I could not have written this book. He was among those bryologists who caringly taught me differences between difficult species on bryological forays. He encouraged me with his charming letters. He provided a book on Michigan mosses just as I arrived in the Upper Peninsula. And now, as I present my own first book on the subject, he has offered numerous helpful comments on an early draft of this book. For repeatedly being my teacher, I thank Dr. Howard Crum.

Dr. Zen Iwatsuki provided the line drawing of *Sphagnum* and several photographs. The *Calliergon* drawing was made by Dr. Yenhung Li.

Janice M. Glime

INTRODUCTION

This book is intended for those who walk the many trails of the Upper Peninsula, Isle Royale, and elsewhere in the Upper Great Lakes, who have an interest in the plants around them, but who have no training in the identification of mosses or their relatives, the liverworts. It is deliberately non-technical in much of its terminology, while introducing its readers to some basic terms that are applied to many of these plants and need to be understood to communicate very clearly the sometimes subtle differences among species. Since books for the beginners are practically non-existent in these fields, the book addresses a wide audience of both children and adults, being sometimes fanciful in the behavior of its participants, but accurate in its descriptions of the plants. More precise terms are often available, but would perhaps burden the reader to the point of dismay, so the serious bryologist is referred to more technical books.

Tetraplodon mnioides growing on dung. (Photo by Zen Iwatsuki)

All my life I have been fascinated by miniature worlds. I could watch miniature train yards for hours, and as a child I never ceased to amuse myself playing with doll houses or miniature Christmas villages. It is therefore not surprising that I chose as my career the study of one of the smallest groups of land plants. This book is a journey in this miniature world through the eyes of one who has always dreamed of being able to walk into that world on its terms — at its size, with its intrigue, joining with its inhabitants.

As we walk among the forest trees just after the snow has disappeared, we can't help but notice the patches of green that signal tiny forests of mosses, sometimes intermixed with the more subtle colors of lichens. But as spring progresses, the elfin world seems to disappear, being supplanted by the giant flowers and ferns that tower over it.

In this book, we shall venture into the land around Lake Superior and across the bogs, ridges, and shorelines of Isle Royale, exploring it as if we are small elves, climbing through the thicket of mosses, examining each for its uniqueness and its place in the scheme of things. We will pause occasionally to examine the lichens, but because our time is short, we will leave that fascinating world for another adventure. Because most of these plants are widespread, we could just as well be in Pennsylvania or Indiana or Minnesota, clambering across the landscape in search of strange adventures.

On all its northern boundary, the Upper Peninsula faces the great Lake Superior, and to the southeast, Lake Michigan. Its western shores are mostly basalt and conglomerates. Its eastern shores are sandstones — red clayey sandstone along the Keweenaw Peninsula and alkaline sandstones at Pictured Rocks National Lakeshore. Its small population of humans is often hidden among the extensive forests of sugar maple, Jack pine, and spruce-fir that mark the transition from the hardwoods to the boreal forest. In that vast forested land, there must be more than 400 species of mosses and liverworts hiding, some waiting to be discovered for the first time.

Two national parks preserve large tracts of natural beauty. The superscripts **IR** and **PR** are used throughout to designate those bryophytes found at Isle Royale and at Pictured Rocks, respectively. Pictured Rocks National Lakeshore occupies 71,397 acres of land that lies along the southern shore of Lake Superior in the north-central section of the Upper Peninsula of Michigan, 46°N latitude, 86°W longitude. Its most spectacular feature is a 20 km (12 miles) shoreline of multicolored sandstone cliffs rising, at places, almost 60 m (200´) above the lake. Its sand dunes, reaching 90 m (300´), are unique in the Upper Peninsula and provide unusual ridge and valley habitats supporting a variety of plant communities.

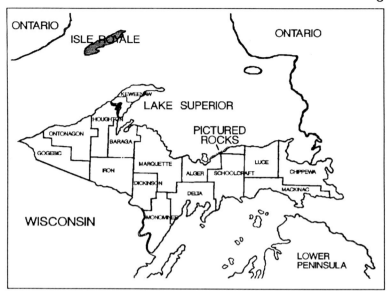

Map of the Upper Peninsula of Michigan and Isle Royale.

Isle Royale National Park is located in the northwestern part of Lake Superior, 48°N latitude, 89°W longitude. The island has over 571,700 acres, is 72 km long and 14 km wide. It consists of a series of parallel ridges, valleys, and islands created as a result of the Precambrian Lake Superior syncline and later glacial erosion. The elevation ranges from 183 m above sea level at the Lake Superior level to almost 427 m above sea level on the ridges. Lack of disturbance and its unique boreal habitats supported the designation of Isle Royale as an International Biosphere Reserve. Due to its geologic configuration and climate, Isle Royale has many species of both higher vascular plants and bryophytes (mosses and liverworts) that are known elsewhere only in the western part of the United States, or farther north.

Most of Isle Royale National Park is covered by coniferous and northern hardwood forests in different stages of succession. Fire plays a major role in the maintenance of poplar and pine forests. The northeastern part of the island has paper birch (*Betula papyrifera*) and quaking aspen (*Populus tremuloides*) on the ridges, with some remaining rock outcrops and areas of young sugar maple (*Acer saccharum*). The valleys and lake shores have balsam fir (*Abies balsamea*) and white or black spruce (*Picea glauca* and *P. mariana*) with white cedar (*Thuja occidentalis*) swamps in the wetter areas.

Cliff Wetmore, a University of Minnesota lichenologist, has found elevated, although not lethal, accumulation of polluting elements, including heavy metals, at some localities of Isle Royale National Park, especially on the ridge tops in the northeastern half of the island. Trees respond often indirectly to the effects of heavy metals and acid rain on the soil and may require a period of several decades to produce measurable symptoms. Bryophytes and lichens, on the other hand, absorb water directly through their photosynthetic surfaces, or "leaves."

Acid rain affects the solubility of metal and nutrient ions in mosses. As a result, acid rain might influence the role of bryophytes in the nutrient cycle of boreal forests. Heavy metals, associated with acid precipitation, can interact with other ions and affect leaching and uptake of nutrients by bryophytes. As a consequence, bryophytes have been used successfully as biomonitors of air pollutants, serving as an early warning of damage to the ecosystem. Joyce Longwith, a former Michigan Tech graduate student, has demonstrated the sensitivity of five boreal bryophytes to pH values of 2.5-3.5, a pH range that occasionally can be found in some Isle Royale precipitation events, but it is more likely in fog. The **pH** is a measure of acidity, with 7 being neutral, and a lower pH means it is more acid. It is therefore important to understand the delicate bryophyte communities so that they can provide us with information that may help us to preserve our forests.

Because this book is intended for field identification, to serve as a beginning for those who are curious, it is by no means complete. It includes those plants I find frequently, or of special interest, and that I feel can be identified with the aid of a photograph and brief description. They are arranged by habitat, though many occur in more than one habitat, and are alphabetical within each habitat. In some areas, you could see all of these habitats in one day's walk.

For nearly every plant I have supplied a previously published common name, that is, one in plain English. However, if you catch me away from this book and ask me the common name of a moss, I can name very few. Scientists do not use common names for these plants, but instead use a Latinized name (scientific name) that is systematically applied to the plants. As you will soon see, some of the common names are silly and cumbersome. Many of them are mere translations of the Latin and are often more difficult to remember than the scientific names. Most American mosses have never been given common names, and the names supplied here are largely based on names published in Europe and Japan. However, scientific names are intimidating at first — they are in a foreign language. Therefore, for some of you, it will be easier to learn the common names first until you become more familiar with the scientific names. Nevertheless, I urge you to learn the scientific names, because then, as stated by my Japanese friend who knew very little English, "you can talk to anyone in the world."

You might want to purchase a 10x hand lens so you can join me in this elfin adventure of viewing the exquisite tiny parts. The lens is designed to help you focus on things that are close to your eyes, so hold it to your eye (just slightly farther than a pair of eye glasses, so you can keep them on if you need them) and hold the lens close to the moss, or stoop down until your nose nearly touches the ground! (It is illegal to pick plants in the National Parks.)

Bryophytes

Everyone knows mosses as delicate plants that spill over rocks, climb trees, and generally live where little competition is offered. Yet, each of these tiny plants has very special requirements. Once their habitats are disturbed, a long time passes before new mosses arrive and grow to sizes large enough to replace the old. This is partly because the plants grow slowly, partly because sources of parent plants to replace them may also be few, and partly because the habitat is no longer suitable for the moss plant to begin its growth. On Isle Royale, where humans have done little to disturb or pollute the pristine habitat, mosses are common.

Mosses can occur in some unusual habitats, such as this shoe.

The mosses and liverworts are known to the professional as **bryophytes**, and the people who study them are **bryologists**. The most similar ones are grouped into one **genus** (pl. **genera**); the members of the genus are termed **species** (pl. **species**), implying that there are detectable differences among these members. Although to the untrained eye mosses may all seem to be the same, you can soon learn to recognize some of the more conspicuous and important ones. Sometimes differences are difficult to detect in the field, so only those species that can be differentiated in the field will be treated here.

To most people, mosses are mosses. But in fact, there are more than 300 kinds in Michigan, and probably nearly as many in Isle Royale National Park or at Pictured Rocks National Lakeshore. This book will serve as an introduction to get you interested in the more visible ones. If you want to learn more about Michigan mosses and liverworts, I recommend the books by Howard Crum, *Mosses of the Great Lakes Forest*, and *Liverworts and Hornworts of Southern Michigan*, published by the Herbarium, University of Michigan, Ann Arbor.

In addition to mosses, the bryophytes include the **liverworts**. Liverworts are of two very different types, the **leafy** ones and the **thallose** ones. Thallose liverworts resemble a green ribbon lying flat on the ground, and most of these have no upright structures. The leafy liverworts at first glance look like mosses, but upon closer examination one finds that they have two rows of leaves that usually lie in one plane, thus causing the plant to appear flat. The leaves are often round, but may be dissected or lobed in various ways, a property that does not occur among the mosses. The most striking difference is in the capsules, which open by splitting into a four-pointed star.

People have traditionally considered the mosses and liverworts to be the first land plants. As we gather new evidence, we find that this may be only partly true. More and more evidence suggests that most of our major land plant groups may have originated on land at about the same time. But two very different possibilities were soon evident. Those plants that had developed lignin, the material that makes trees woody and permits your broccoli to grow upright, were able to attain large sizes, contributing to huge plants early in their evolution. Those that lacked lignin, the bryophytes, were unable to attain any significant size during their entire evolutionary history.

Lacking lignin, the bryophytes were unable to develop a sophisticated system for internal transport of water (vascular tissue), another factor that has helped to keep them small. Instead, they have developed external water movement by providing a series of capillary spaces and operating much like a sponge or wick. They often retain water by growing in dense cushions, and the moss that ventures to grow faster and exceed the cushion is soon slowed by lack of water at its growing tip, thus permitting the other stems to catch up with it and retain the cushion shape. Only a few mosses, such as the common hairy cap (*Polytrichum*), have developed an internal system sufficient to supply most of their water and nutrient needs from the soil.

Life Cycle

For reasons we do not fully understand, a second phenomenon seems to be coupled with the development of lignified vascular tissue. Life cycle stages that have lignin and lignified vascular tissue are the life cycle stages that possess two sets of chromosomes, *i.e.*, they belong to the generation that results from sexual reproduction. For this to make any sense, we must realize that plants (as well as animals) have a life cycle stage with duplicate chromosomes ($2n$), and that they have a division process called **meiosis** that divides these two sets and places them into different cells, reducing the number of chromosome sets back to one. In animals, we recognize the resulting cells after meiosis as **gametes**, but in plants, there is a delay after meiosis before gametes form, and the plant continues to grow and carry out activities with only one set of chromosomes. This is especially true in mosses, where a whole leafy plant develops during this stage with one set of chromosomes, the **gametophyte**.

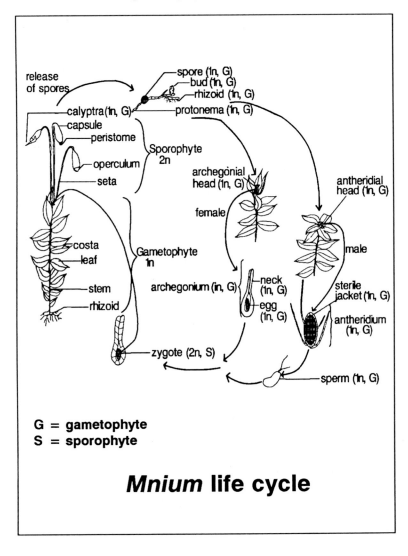

G = **gametophyte**
S = **sporophyte**

Mnium life cycle

When this leafy **gametophyte** is mature, it reproduces by sexual reproduction, where gametes unite, resulting in a **sporophyte** having two sets of chromosomes again. This is the stage we recognize as the stalk and capsule of the moss. However, in the mosses, no lignified vascular tissue is present in this sporophyte, as it is in all the other land plant groups. It is within this capsule that meiosis occurs, making spores that can then grow into the leafy plant once more. We will examine the fascinating structures of the moss more carefully to elaborate on this scheme of alternating generations.

When you examine your first moss, you may be surprised to see that its tiny, delicate leaves can have a rib down the middle (**costa**) and sometimes even a **border** of narrow cells. It has only been a few years since we found out that the costa is able to transport water in many of these mosses, making them more like the vascular plants than we ever imagined. The stem is many cells thick and often has large, thin-walled outer cells surrounding cells of a much smaller diameter in the center. If we could examine these vertically, we would find that these smaller cells are actually longer and often function to move water through the stem to the leaves. But in most mosses, the water moves much more rapidly on the outside. Watch it for yourself by dropping a few drops of water on a dry moss, and you will find a very lively moss "coming to life."

Leaf of *Rhizomnium* showing a distinct border and costa. (Photo by Zen Iwatsuki)

Mosses and liverworts lack roots, but they are often anchored by tiny structures called **rhizoids** that resemble roots and function mostly in anchorage. These structures are but one cell wide, although they are several cells in length. The most common position of the rhizoids is at the base of the stem, but they extend along the stem in some species, affording them additional points of attachment or aiding in water movement and perhaps in its retention.

Fontinalis hypnoides showing rhizoids.

Mosses can survive long dry periods much better than most higher vascular plants, and when they are dry, tiny bits of leaves or stems often travel quite some distance. In my garden room, my pet box turtle would carry my liverworts all over the room without ever knowing he had any passengers. These tiny bits are very important to the bryophytes because they are able to grow into entire new plants!

Once a year, most mosses have sexual reproduction, a process that requires at least some water, so the event is most commonly timed to occur during the season when water is most likely to be available. Tiny male sperm are produced in sacs (**antheridia**) that are sometimes located in special cups (see *Polytrichum*) or platforms on top of a plant, where they are readily splashed, hopefully onto a female plant. In other cases, both sexes are on the same plant, although often they are not both there at the same time, thus still requiring a partner from a different plant. The sperm are capable of swimming and look a little bit like a microscopic animal with their two flagella.

Fontinalis hypnoides with antheridia (upper) and *Fontinalis dalecarlica* with archegonia (lower).

These tiny sperm can swim in water films on the plant and down the neck of the female organ, the **archegonium**. If you look carefully among the leaves at the very tip of upright mosses or along the stems of horizontal ones, you may see these tiny flasks in a small cluster. There the sperm joins an egg, and thus the stalk (**seta**) can begin its development.

Stalks and capsules of *Pohlia* (upper) and calyptra on a mature capsule of *Plagiopus oederiana* (lower).

At first, this stalk is hidden beneath a covering, or cap, called a **calyptra** that develops from the archegonium. As the stalk grows, it stretches the calyptra, and finally the calyptra perches atop the slender seta like a stocking cap on the end of a flag pole. At last, the **capsule** expands and as it matures it produces tiny **spores** inside. These spores can make new plants, but not until they venture away from their parents.

Escape from the parent is facilitated by one of the most beautiful features of a moss, the set of teeth (**peristome**) at the top of the capsule. These teeth can be long and slender, and they often have tiny shelves on one side toward the center of the capsule. Spores can lodge here, and as the moisture changes, the teeth move about, permitting the spores to leave and travel on a puff of wind or drift to the ground.

Peristome of *Bartramia pomiformis*. (Photo by Zen Iwatsuki)

In some mosses, the spore will grow right away if it reaches a moist surface, but in others it may wait a long time, perhaps resting in the soil until something brings it to the surface. When the spore gets both light and moisture, it swells, then sends out a long, green thread (**protonema**, pl. **protonemata**) as if it were an alga. This thread can cover several centimeters (1" or more) of ground before it changes its behavior and makes an upright leafy plant. But one protonema will usually make lots of leafy plants, and thus this moss will not be alone, but will be part of a mat or cushion or other arrangement of neighbors, helping each other to escape the drying wind and burning sun.

Now the plants are ready to live there for several months to many years, using the sunlight for energy when they are wet, but resting when they are dry. And each kind of moss has its own special preferences that will make your adventure through this book so intriguing.

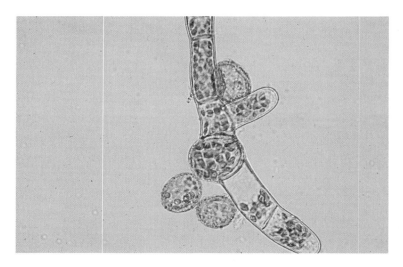

Germination of spores, showing emerging protonema (upper), and mass of moss protonemata before the leafy shoots develop (lower).

Moss Neighbors

In 1850, Foster and Whitney wrote of Isle Royale, "The shores are lined with dense but dwarfed forests of cedar and spruce, with their branches interlocking and wreathed with long and drooping festoons of moss." Since pendent species of mosses do not occur on the island, or anywhere in the region, I presume they were referring to the **lichens**, especially **old man's beard (*Usnea longissima*)**, which is often confused with mosses. As you travel among the mosses, you will surely encounter many lichens. Humans often confuse these strange plants with mosses, and some of them, such as **reindeer moss**, even bear moss names.

Old man's beard hanging from boreal spruce trees.

Reindeer mosses include such lichens as *Cladonia rangiferina* (upper left), *Cladina uncialis* (lower left), and *Cladina mitis* (lower right).

Because of a unique partnership (**symbiosis**), this unusual group is able to live in places where no other organisms can. An example is *Xanthoria elegans* living on exposed rock faces of the Lake Superior shore. Although lichens appear to be one organism, and indeed can be identified as unique beings, they are a combination of a fungus living with a green alga or blue-green bacterium (**Cyanobacteria**, formerly known as blue-green algae). The fungus provides the structure and presumably maintains the water level and shields the alga from light during dry periods. The algae and blue-green bacteria, both being photosynthetic organisms, provide the energy for the system. There are probably other benefits in at least some cases, such as anchorage and contribution of hormones from the fungi, nutrient trapping and nutrient gathering by the fungi, and contribution of vitamins from the photosynthetic partner. These closely attuned organisms no longer resemble their free-living counterparts (if such free-living members even exist).

Lichens are most easily recognized from mosses by their lack of bright green color (with few exceptions), lack of leaves, and more leathery texture. Some have brilliant red caps that store fungal spores (**British soldier lichen**, *Cladonia cristatella*); others resemble tiny goblets (**goblet lichen**, *Cladonia chlorophaea*). They grow almost anywhere that isn't hot and moist at the same time, but in spite of their algal partner, few of them grow under water.

The British soldier or match head lichen (*Cladonia cristatella*) (left) and the goblet lichen (*Cladonia chlorophaea*) are common on decaying wood and soil.

The lung lichen (*Lobaria pulmonaria*) is common on trees at Estivant Pines near Copper Harbor.

The most common ones are the broad, thalloid ones that grow on the trees and rocks and that have been used to

18

assess acid rain damage, including the **lung lichen** (*Lobaria pulmonaria*), the **wrinkled shield lichen** (*Flavoparmelia caperata*), **rock tripe** (*Umbilicaria*), **dog lichen** (*Peltigera canina*) and **ring lichen** (*Arctoparmelia centrifuga*).

The wrinkled shield lichen (*Flavoparmelia caperata*) is a conspicuous epiphyte throughout the UP.

George Washington reputedly fed rock tripe (*Umbilicaria*) to his troops in a broth of water, providing a hot soup when the winter food supplies were low, but there seems to be no documentation that such an event ever actually took place.

On forest soil banks among mosses one can see the highly tarnished teeth-like apothecia of the **dog lichen** (*Peltigera canina*) protrudong at the tips of upturned lobes, resembling the teeth of a dog. This character led the European herbalists to presume it had such uses as a cure for rabies, and some folks put it in their shoes to prevent dogs from biting them! By the same token, since dogs get rabies, making them hydrophobic, it was used as a cure for hydrophobia.

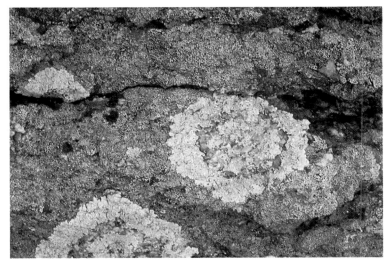

The yellow-green concentric rings of **ring lichen** (*Arctoparmelia centrifuga*) resemble those of a target, resulting from recolonization of the center of the thallus after the older portions have died and disappeared.

Mat of *Sphagnum* surrounding a lake near Perrault Lake in Houghton County, Michigan. This type of mat has traditionally been called a bog in this country, but it is in fact a poor fen, deriving most of its moisture from ground water.

BOGS AND FENS

Bogs and fens have a mystique all their own. Early morning haze from the condensing moisture gives the spruce and larch trees, draped with curtains of lichens, an eerie look. Bodies of men have been recovered from spongy, wet ground of bogs and fens in Denmark, perfectly preserved for hundreds of years in the acid water deep within the blanket of bog mosses. These men were not decayed, and the ropes used for hanging them still encircled their necks! There has been much speculation about what rituals must have surrounded these deaths, because the men had peaceful smiles on their faces and full meals in their bellies.

Bogs and fens are one of the few places where mosses dominate the landscape. To use this book it is not really necessary for you to distinguish bogs from poor fens. **Bogs** are wet habitats that use rainwater as their only source of water and nutrients, whereas **fens** use water from streams or ground runoff as well. This means that bogs must be raised slightly above the immediately surrounding landscape as well as above the groundwater, while at the same time maintaining a high water content. The high water content is maintained by

peat moss (*Sphagnum*). The peat moss acts like a wick, carrying water from a lake or pond upward into the mound of moss plants. Other plants, like the hairy cap moss, pitcher plants, sundews, and cranberries, are able to maintain a high moisture level because they live among the very wet peat moss. These peat mats often extend over the water as floating mats, underlain by roots and stems of shrubs, and if you jump on them, you will feel the "ground" beneath you quake. However, be careful! These floating mats can often have thin spots where you could fall through and be unable to find your way out of the dark waters beneath.

Bogs are generally acid, whereas fens include the acidic **poor fens** with *Sphagnum* mats, **intermediate fens** of nearly neutral pH and moderate nutrients, and **rich fens** that are generally alkaline (high pH) and rich in nutrients, especially calcium and magnesium, and are dominated by sedges and brown feather mosses. While all three types of fens can have *Sphagnum*, it is generally dominant only in the poor fen, with *Drepanocladus* and sedges being more common in rich fens.

Walk here with great caution! Bogs and fens are delicate ecosystems. Not only can they be dangerous, having areas of thin mat where the unsuspecting explorer can disappear forever under the mat, but they are easily destroyed. The weight or your feet breaks the tiny branches and submerses the plants under the less acid lake water where the plants often die. After a few years of berry pickers making a path across the mat, the path becomes a water trail. The invading waters soon destroy wider and wider areas of the mat as they bring water of a different pH into the mat.

Peat Mosses

Sphagnum (Sfag′ num) — **peat moss**

dense branches
at tip

branches usually
in groups

long stems

older parts
brown

Sphagnum, actual size, drawn by Zen Iwatsuki.

There are more than twenty species of *Sphagnum* in Michigan. These have many unusual abilities that make the genus very important. Its ability to hold large quantities of water (up to 30 times its own dry weight) makes it useful in shipping live fish and plants. Most of the orchids you can buy to grow at home have peat moss wrapped around their roots. It is often used in flower beds to help keep the weeds from growing. One boot manufacturer in Europe uses peat moss to cushion the soles of hiking boots!

In addition to its moisture-holding qualities, peat moss contributes acid to its surroundings. It does this by making a trade. It takes things like calcium and magnesium ions out of the water and replaces them with hydrogen ions. It is the hydrogen ions that make the bog and fen water acid.

The acid and absorptive qualities of peat moss made it very useful in World War I. The acid keeps bacteria from growing, so it was used to make bandages for the soldiers. Many wounds actually healed better with the peat moss bandage than with sterile white bandages because there were fewer problems with infection.

Apparently because of the lack of oxygen in the deep layers of peat, the process of decay and nutrient release in a bog and acid fen is slow. As a result, plants in those habitats must be able to survive with low nutrient concentrations.

Sphagnum leaf showing photosynthetic cells and hyaline cells. Fluorescent microscopy makes the chloroplasts appear red and shows the flavonoid compounds in the cell walls.

One can recognize the peat mosses by their dense heads of branches (**capitulum**) at the tops of the plants. These heads are usually pale green due to the large water-holding cells, but some species are olive-brown, brown, or yellow, and many have bright red pigments, especially in the fall when nights are cold and days are sunny. Under the microscope, one can see that the moss has an unusual network of cells, with long, colorless (**hyaline**), water-holding cells surrounded by about six narrow, green (**photosynthetic**) cells. These water-holding cells often have tiny algae or microscopic animals living in them. It is these water-storage cells that make the moss so important for shipping fish and plants.

The capsules of *Sphagnum* are soft spheres, and the stalks that support them are watery and often short-lived, so seeing these interesting capsules still attached is a special treat. The capsule has a tiny lid (**operculum**) on top, and as the spores mature, gases build up inside the capsule, putting pressure on the operculum. As the capsule begins to dry, more pressure is exerted on the shrinking capsule and ultimately the operculum is blown off explosively. You can hear the capsules popping their lids, and if you forget about this and place them under your reading lamp, they will begin exploding as they dry out. When you walk through a patch of these on a sunny day, listen carefully!

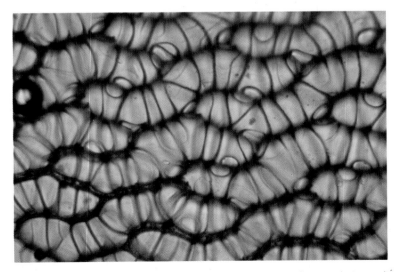

Stained cells of *Sphagnum* showing two cell types and pores that permit water to enter and be stored.

Open capsules of *Sphagnum rubellum* showing the lack of teeth.

Sphagnum magellanicum PR (madge ell lann´ ih kum) — **midway peat moss** — is one of the most common *Sphagnum* species, often displaying a brilliant crimson color. It is one of the large *Sphagnum* species, having a somewhat fleshy appearance and relatively large leaves. It is especially red in autumn, but likely to be pale when the snow first melts and uncovers it in spring. Low nitrogen content, high light intensities, and cold temperatures stimulate the production of the red pigments.

26

Sphagnum papillosum[IR,PR] (pap pill loh′ sum) — **papillose peat moss** — is among the close relatives of *S. magellanicum.* It can be recognized in the field by its yellow to brown color and large size, like that of *S. magellanicum.* It often occurs with *S. magellanicum,* but is somewhat less common.

Sphagnum squarrosum[IR,PR] (square oh′ sum) — **spreading leaved peat moss** — likewise is a related large moss, but it is never red or olive colored, and it is usually in a shaded area. It grows most often in the forest, especially forests with spruce (*Picea*) or hemlock (*Tsuga*), and in white cedar (*Thuja*) forests, but not in bogs or fens. Its name refers to the leaf, which goes up along the stem, then spreads widely away from the stem by having a bend in the leaf itself.

Sphagnum fallax[IR,PR] (fal′ lax) — **pointed white moss** — has paired short branches between each of the main branches of the capitulum and is light green to yellow-brown. It and all the following *Sphagnum* species have smaller leaves than the preceding species.

Sphagnum fuscum[IR] (fuss′ kuhm) — **rusty peat moss** — is brown with crowded small, rounded capitula (heads), and can be found on the tops of the numerous small mounds called **hummocks**. You'll remember the hummocks because they can make walking in a bog or fen quite tiring. These mosses prefer the more acid older open bogs.

Sphagnum wulfianum[PR] (wolf ee ay´ num) — **ball white moss** — has a very dense capitulum more than 1 cm (.5") across and has a dull look, partly because its branch leaves are curved outward. Its stems are dark and brittle. This moss prefers bog forests, often on hummocks of old decaying logs.

Sphagnum girgensohnii[IR,PR] (guhr genn soh´ nee eye) — **white-toothed peat moss** — is a large *Sphagnum* with very long branches, compared to other *Sphagnum* species. Like *S. wulfianum*, it prefers bog forests, but also is common in *Thuja* forests, preferring at least a moderate nutrient supply.

Sphagnum cuspidatum[PR] (kuss pih day′ tuhm) — **drowned kittens** — grows submersed and indeed has the look of the stuck together fur of a wet kitten. The young pendent branches do not occur in pairs and do not hide the stems as they do in most *Sphagnum* species, separating it from its look-alike *S. recurvum*.

Sphagnum rubellum[PR] (rew bell′ lum) — **red bog moss** — is a common red species with small leaves and pompon-like capitulum (head). This is a distinctly acid species on the tops of hummocks.

Intermediate fens

Like a bog, the intermediate fen lacks a forest canopy. However, instead of *Sphagnum*, there are other species of mosses. The shrubs are different too; the bog laurel, bog rosemary, and cranberries are gone.

Aulacomnium palustre[IR,PR] (All uh kohm' nee uhm   pah luss' tree) — **ribbed bog moss**

The ribbed bog moss *Aulacomnium palustre* (top) with propagules (bottom, photo by Zen Iwatsuki).

As you crawl on your hands and knees, observing the world through a lens, you may suddenly fear for your life as you encounter the medieval-looking structures resembling spiny clubs. These strange-looking objects are the vegetative reproductive stem tips of *Aulacomnium palustre*. Each stalk has a series of tiny, thickened and highly modified leaves (**propagules**) that are able to break away and start new plants. In slightly drier hummocks (mounds) of bogs and poor or intermediate fens, you may find these light or yellow-green clumps. The leaves have a strong rib (**costa**), and when *Aulacomnium* is fresh and moist, its leaves are erect and **keeled** (folded along the middle like a boat). When the moss gets dry, its leaves become contorted. The leaves look dull because of tiny bumps (**papillae**) on the cells. Usually this moss has numerous reddish-brown threads (**tomentum**) near its base, helping to anchor it and to increase **capillary** water movement. Because of these threads, they will have more small spaces where water will move quickly to fill the space.

Tomenthypnum nitens[IR] (Toe ment hip′ nuhm nigh′ tens) — **shining feather moss**

As you wander about in these more alkaline (calcareous) fens, where more nutrients are available than in bogs and poor fens, *Tomenthypnum nitens* may stand erect from the mounds (**hummocks**), often with *Aulacomnium palustre*. Since it is more common in calcareous habitats, it usually occurs where there is little *Sphagnum*, except *S. warnstorfii*.

The large size and golden color of *Tomenthypnum nitens* are distinctive, but even more distinctive are the dark brown threads **(tomentum)** that cover the blackish stems on one side and facilitate water movement. The leaves are straight, strongly folded like an accordion **(plicate)**, and have a rib (costa) that ends three quarters up the leaf. The leaves have sharply pointed tips and give the branches the appearance of a fine artist's brush.

Rich fens

These nutrient-rich habitats are characterized by the brown feather mosses, sometimes called sickle mosses, that can help to fill in lake margins and make it possible for other plants to grow above the water level where the pH is not quite so high. Hummocks of sedges, interspersed with small pools, often make walking difficult.

***Drepanocladus revolvens*[IR]** (Dreh pan ah' clah duss   ree vahl' venns) — **rusty claw-leaved feather moss**

Although the name implies that this is a rusty-colored moss, I think of it as a beautiful purplish bronze moss with leaves shaped like the blade of a sickle. A costa (rib) runs the extent of the leaf, thus enabling us to separate it from similar mosses in the genus *Hypnum*. It is not unusual to find this species growing together with *Drepanocladus vernicosus* in the same fen, usually differing from it by its color and lack of waves in the leaves.

Drepanocladus vernicosus[IR] (Dreh pan ah' clah duss   vern ih koh' suss) — **green feather moss**, or **sickle moss**

If you rest on a hummock of sedges, you might encounter branches of miniature scythe or sickle blades, all pointed in a single direction as if neatly hung up for the season. These blades most likely are the long, curved leaves of *Drepanocladus*. Among these, *Drepanocladus vernicosus* usually prefers sites that have more nutrients and less acidity than those of *Sphagnum*. *Drepanocladus vernicosus* has slightly wavy leaves with a long costa (rib) and is usually a simple green. It grows in rich fens and can help to provide a suitable moist, but not submersed, habitat for shrubs. After these shrubs become established, the *Drepanocladus* and shrubs provide a safe habitat above the alkaline water where *Sphagnum* can survive. If we dig down to the lower peat layers of many older bogs, we can find layers of woody shrubs and *Drepanocladus* that tell us the plant successional history of the area, and we can understand the importance of these plants in changing the habitat.

Scorpidium scorpioides[R] (Score pihd' ee uhm   score pee oy' dees) — **scorpion feather moss**

If you wander too far and step into a water-filled channel or pool, you may encounter the scorpion-like stems and branches of *Scorpidium scorpioides*. This beautiful moss often has crimson leaves that are relatively large and wide. It grows in rich fens and is usually submerged in water where its dark color against the dark water can make it relatively inconspicuous. If you do see it, it may look like a pile of worms just under the surface! Look for it among sedges and shrubs in the water of very nutrient-rich habitats where the pH is relatively high (alkaline). It will not grow in a *Sphagnum* mat, but may have *Sphagnum* neighbors nearby.

As the name implies, the leaves and tips of *Scorpidium scorpioides* curve to one side like the tail of a scorpion. The leaves are 2-3 mm (0.1") long and often wrinkled or wavy. The costa is very short and double, or even absent, and you are not likely to see it. The coloring of the plant varies considerably from green to yellow to brown to crimson.

MARSHLAND

If you like soggy feet, marshlands offer their own unique bryophytes, where the high nutrient content and usually high pH favor species quite different from those of bogs. Instead of a mat of *Sphagnum*, you will be confronted with tall, thin, grass-like vegetation, usually in standing water. Down at the bases of cattails, you can find mosses climbing the stems; between the stems there are often several species poking their tips above the water, some of which are quite unusual.

Cattail marsh, demonstrating the vertical type of vegetation that characterizes many marshlands.

This is not a stable habitat because the tall cattails and sedges die each winter, necessitating new colonization by anything not living in the water itself. Furthermore, the light is greatly diminished by these towering marshland neighbors.

Calliergon[R] (Cal lee uhr′ gahn) — **exquisite feathermoss**

As you slosh among the reeds, you can find *Calliergon* species submerged and emergent among the cattails and sedges of wet meadows, woodland pools, and fens. In the North, they are sometimes important in controlling the water from spring runoff, absorbing it like a giant sponge.

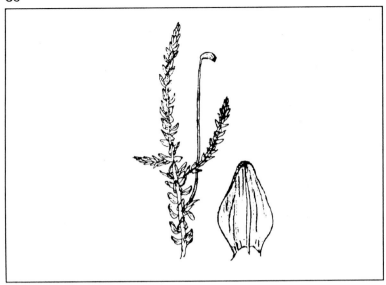

The name of the moss means "pretty work" and refers to its elegance of construction. Most of these plants are light or yellowish green and somewhat shiny. Unlike their close relative, *Calliergonella*, they have a costa (rib) in their leaves. These leaves are concave (like a bowl) and the tip is blunt and especially concave, giving the appearance of a hood. These leaves are held closely to the stem, giving the plant a tidy appearance — an elegant construction.

Calliergonella cuspidata (Cal ee uhr gahn ell′ luh   cuss pih day′ tah) — **pointed bog feather moss**

Emerging from the water surface of rich fens and alkaline sedge meadows *Calliergonella cuspidata* does somewhat resemble a loose feather, having shiny green or yellowish spikes with branches and a pointed tip. The leaves, like those of *Calliergon*, are relatively blunt-tipped, concave, and hooded; the bases have large, transparent cells that can sometimes be seen even with a hand lens, but they lack a costa.

Calliergonella cuspidata

Campylium[IR,PR] (Cam pill′ ee uhm) — **golden creeping moss**

As you wander through the drier parts of the marsh where shrubby cinquefoil grows, you may see tiny golden threads that appear to have fine bristles on them. These are most likely species of *Campylium*.

Several species of *Campylium* occur in damp habitats of fens, wet meadows, and stream banks. The easiest way to recognize them is by their spreading leaves that seem to bend somewhat abruptly and point at 90° from the stem. These mosses are usually light green or yellowish and the leaves are narrow toward the tips, ending as a long point. Some of the species have a costa (rib) in the leaf and others do not.

Climacium dendroides[IR,PR] (Clai may′ see uhm   den droi′ dees) — **tree moss**

When you reach the stream bank or tramp through wet forested areas, you may think you have become a true giant looking down on a forest canopy. The elegant *Climacium dendroides* looks like a miniature tree (dendroid); its relative in Japan has been used commonly in dish gardens to represent tall trees. It was at one time sold in both England and Boston to adorn ladies' hats, where its appearance was improved little by the addition of a green dye. In Japan, it is made into ornamental flowers to be floated in water. Another *Climacium* species is made into wreaths and crosses and sold at Christmas in the United States.

It is distinguished in this part of the world by having a stem nearly barren of leaves, but topped by a cluster of branches that give it a tree-like appearance. If one pulls it up, a long **rhizome** (horizontal stem) will follow, and it is likely that other upright plants will be attached. The leaves are distinctive because they are **plicate**, that is, folded like an accordion. Their tips are blunt and they have a long costa.

Hypnum lindbergii[R] (Hip´ nuhm   lind berg´ ee ai) — **creeping moss**

As you wander along the stream banks and among the wet meadows in the spring, you will encounter *Hypnum lindbergii*, but later in the season grasses and other plants obscure it from view. In such habitats, it is undoubtedly important in preventing springtime erosion. *Hypnum* has been used to stuff pillows and induce sleep in the Fiji Islands.

Its shiny, pale green color, leaves 2-3 mm (0.1") across, curved leaf tips, and flattened appearance make this moss relatively easy to recognize. The leaves lack a costa (rib), which one will find in the similar *Drepanocladus* species, so it should be easy to distinguish in these habitats.

CEDAR SWAMPS

Hiking through the white cedar (*Thuja*) swamp won't be an easy chore; the swamp has a tangle of fallen logs in numerous stages of decay. As a result of this tangle, multitudinous niches with widely differing conditions provide a site rich in bryophytes. Wet soil, pools, decaying logs, and shade make the habitat ideal for many species and lots of bryophyte cover. Many of these species are also found in the spruce and Jack pine forests (discussed later), including *Thuidium delicatulum, Ptilium crista-castrensis, Hylocomium splendens, Pleurozium schreberi, Tetraphis pellucida,* and *Eurhynchium pulchellum,* as well as some from marshland, such as *Calliergon.* But some species are difficult to find elsewhere.

Rhizomnium magnifolium[IR] (Rai zoh nai´ uhm   magg´ nih foe´ lee uhm) — **large-leaved mnium**

Perched atop logs and on decaying wood above the summer water table, you'll find a plant with such large leaves that at first glance it seems certain to be a small vascular plant. This moss, *Rhizomnium magnifolium,* can also be found in bogs, along streams, and in swampy woods. Because of its large size and beauty, it is used in Japanese moss gardens.

The large (6-7 mm long, 0.2"), dark green leaves have a conspicuous border and conspicuous costa that are often reddish. The stem is usually unbranched, and the upper leaves are larger than the lower ones.

Trichocolea tomentella[PR] (Trih ko coh′ lee uh   toe men tell′ luh) — **common down liverwort**

The largest and most beautiful leafy liverwort is seen only in damp forests such as white cedar (*Thuja*) swamps. It is covered with long hairs, resulting from fine divisions of the leaves, thus giving it the specific name of *tomentella* (having little hairs) and the common name of **down**. It somewhat resembles species of the liverwort *Ptilidium*, but it is larger and pale green or whitened by the hairs.

STREAMS AND LAKES

Aquatic mosses occur in the deepest water known for any freshwater plant. They are the dirge of every fisherman who gets them tangled in a fishing hook. They likewise occur in some of the fastest water of streams, where higher vascular plants cannot survive the constant abrasion and physical stress. Here they harbor numerous aquatic insects that provide food for fish. Their greatest enemies are high temperatures and high nutrient levels, so they serve well as monitors of heated or chemical-rich effluents. Their ability to harbor heavy metals for long periods of time, relative to higher vascular plants, makes them useful for both clean-up and recording of heavy metal spills.

Bryum pseudotriquetrum[IR,PR] (Bry′ uhm　su doh try kee′ trum) — **tall clustered thread moss**, or **marsh bryum**

At the margins of ponds, shallow lake shores, stream banks, fens, and forest pools, this distinctive *Bryum* can abound. It is unusual in its **decurrent** leaf bases (the base extends downward along the stem) and relatively tall stem, sometimes reaching more than 2 cm (1″) in height. Its olive to dark green color and strong border help to distinguish this aquatic species. It can be planted in the bottom of an aquarium to form an attractive small forest that will reach several centimeters in height.

Chiloscyphus polyanthos* var. *rivularis[PR] (kai loh sai′ fuss pah lee ann′ thohss　var. rihv yu lair′ iss) — **square-leaved liverwort**

As you wade in a small stream to cool your feet, you may find *Chiloscyphus polyanthos* var. *polyanthos*. This is the most common leafy liverwort on the rocks in our streams, occupying even small pebbles only the size of a robin egg. The leaves are round and their bases slightly extend down the stem (**decurrent**), with deeply bilobed underleaves possessing rhizoids, resembling small brown tufts or pompons, at their bases.

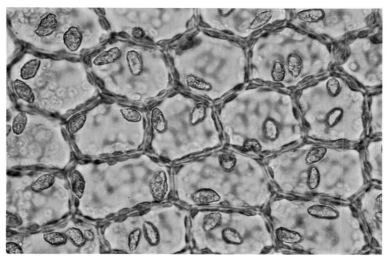

Cells of *Chiloscyphus polyanthos* under the microscope. (Photo by Zen Iwatsuki)

Fissidens bryoides (Fiss′ ih dens bry oy′ dees) — **common flat fork moss**, or **small phoenix tail moss**

While you are turning over rocks to find caddis flies and mayflies, you may be amazed by a microscopic forest of tiny, flat mosses with miniature capsules. *Fissidens bryoides* in our area most commonly grows submersed, especially on the red Jacobsville sandstone, but also on other types of rocks, sometimes among the bases of the large ,aquatic moss, *Fontinalis*. It can occur on damp rocks above the water line, but seemingly does not do so in the Upper Peninsula.

In spite of its small size, its two rows of leaves still have a strong costa. The most unusual feature is a pocket on each leaf into which the next leaf fits. The capsule is on a seta of only a few millimeters in length and is itself smaller than the head of a straight pin and short cylindrical or nearly round. It is striking because of the bright orange color of its teeth.

Fissidens grandifrons (Fiss′ ih dens   gran′ dih frahns) — **narrow-leaved phoenix moss**

When you reach a waterfall, a dark green moss clings tightly to the rocks, withstanding all the force of a spring melt runoff. The beautiful long fronds have reminded the Chinese of the feathers of the lovely phoenix, a bird often portrayed in oriental art. By far the largest of the *Fissidens* species, this one has leaves more than one cell thick, thus giving it a stiff appearance and feel. Since it is one of the few submersed aquatic bryophytes to be found on calcareous rock, it is supposed that it gets its carbon dioxide from the air in the turbulent water; the high pH does not permit carbon dioxide to remain free in the water. It is rare in the Upper Peninsula, but occurs abundantly at the Jacobs Creek falls in Keweenaw County.

Fontinalis[IR,PR] (Fon′ tin al′ iss) — **brook moss**

In other clean, more acid fast-water streams, you will find *Fontinalis* species holding tenaciously to the rocks. If you try to dry it, you will be displeased with its oily, almost fishy odor. *Fontinalis* members lack a costa in their leaves, separating them from many of the other aquatic mosses, especially from *Leptodictyum riparium*, a good *Fontinalis* look-a-like.

Fontinalis antipyretica (Fon′ tin al′ iss   ann tie pie reh′ tih kah) — **fire preventer moss**

As its name *antipyretica* implies, this moss was used against fire in Nordic countries by placing it between chimneys and the timbers of houses as an insulator.

The species *Fontinalis antipyretica* tends to grow in relatively quiet water of streams. Its keeled leaves often harbor tiny midge larvae and other aquatic insects. Like many of the aquatic mosses, these are large mosses, and the common variety **gigantea** has the largest leaves of any moss in this region. *Fontinalis antipyretica* is the only common local species of submersed moss with strongly keeled leaves.

Fontinalis novae-angliae[IR] (Fon′ tin al′ iss   noh vee ang′ lee ai) — **New England brook moss**

   On Isle Royale, you may find long, dangling, brownish mosses hanging from shrubs at the edge of Lake Richie or other lakes during the summer.   These will probably be *Fontinalis novae-angliae*, waiting for the water level to rise again to submerse them so they can resume growth.   This species also grows in relatively fast water of streams, where its numerous **rhizoids**, resembling small, brown roots, cement themselves to the rock and keep it in place.   It will experience periods out of water and thus must remain dormant until it is once again submersed. It produces horizontal stems, enabling it to spread over rock surfaces, and these have been used to help hold rocks in small diversion dams of streams.

   This moss can be used in aquaria as long as the water is cold. It prefers temperatures around 10°C (about 50°F).

Fontinalis hypnoides[IR,PR] (Fon′ tin al′ iss   hip noy′ dees) — **river moss**

   A highly variable species, this moss lives in moderate flow of streams.   It tends to have reddish stems and small, flat, lance-shaped leaves with no rib (see introduction).   There is a larger variety, var. *duriaei*, that usually dies back during the summer and re-grows rapidly during the winter and following spring.   This variety has somewhat concave broad leaves that may even appear to be slightly keeled during certain seasons.

Although the two varieties converge in the South, they are quite distinct in the Upper Peninsula.

Hygrohypnum ochraceum (Hai grow hip' nuhm   oh kray' see uhm) — **yellow mountain-rill feather moss**

In Utah, one can see a bird called the water ouzel diving into the water of streams and carrying back this moss to construct its nest. This moss likewise grows on rocks in small streams of Michigan and closely resembles *Platyhypnidium riparioides*. It differs from that species in having a layer of transparent cells on the outside of the stem. If you pull down on a leaf, some of these cells will usually come with the leaf.

Nest made of *Hygrohypnum ochraceum* and *Hygroamblystegium.*

Hygroamblystegium[IR,PR] (Hai grow am blih stee' gee uhm) — **brookside feather moss**

I always know a member of *Hygroamblystegium* because it feels and looks like a scouring pad for cleaning pots and pans. Its wiry stems form a mat several centimeters (1" or more) thick, and the battered leaves soon lose their soft tissues, especially in the summer after the weather becomes warm and growth ceases. The unique feature of this genus of moss is that when the soft tissue of the leaf is eroded away, the strong costa remains, so that one can see what appear to be tiny bristles projecting from all the stems. This genus is the only common one in our area that has this feature.

The leaves are slightly concave and tear-drop to long tear-drop shaped (**ovate** to **short lanceolate**), bright green in winter and early spring, spreading somewhat from the stem, and appearing to be somewhat rigid. Especially in late winter, this moss harbors numerous immature aquatic insects, particularly midge larvae. The dense mat of stems makes a habitat that is nearly as quiet as a pool.

Leptodictyum riparium[IR,PR] (Lep toh dick′ tee uhm rye pair′ ee uhm) — **willow moss**, or **waterside feather moss**

When I was hunting for *Fontinalis* in New Mexico, my students returned from a scouting trip very excited. On examining their find by the light of a kerosene lamp, I had to break the news to them that even in the dim light I was fairly certain I could see a costa, making their exciting find *Leptodictyum riparium* instead. This mistake in identification occurs often, but *Leptodictyum* has flat leaves with a costa and they are generally in one plane, appearing to arise on only two sides of the stem but actually twisting around the stem to lie there, whereas *Fontinalis* seldom has flat leaves, has no costa, and has leaves arising on three sides of the stem, clearly not in one plane.

Platyhypnidium riparioides[IR,PR] (Plah tee hip nih′ dee uhm rih pair′ ee oi′ dees) — **beaked water moss**

This worldwide species, like most aquatic mosses, houses innumerable insects in streams. It usually occurs on rocks that are only submersed for part of the year, and often it lives in the splash zone at the edge of the stream.

This species is somewhat more successful than *Fontinalis* in aquaria, but it is more compact and less beautiful.

Platyhypnidium riparioides has a rib that extends most of the way to the leaf tip, and its leaves are concave, ovate, and have a short point at the tip. This moss most closely resembles some forms of *Brachythecium rivulare*, but differs at the base of the leaf. In the latter species, the base extends noticeably down the stem, whereas in *Platyhypnidium,* this is not the case.

Riccia fluitans (Rick′ see ah   flew′ ih tans) — **floating crystalwort**, or **forked ribbon plant**

Floating just beneath the surface (*fluitans*) at the edge of ponds, often just under the duckweed, one can find thin green ribbons forking in repeated Y shapes and intertwined in masses. This amphibious liverwort is just as healthy on soil as in the water, but on soil it becomes a broad blade (thallus) that develops thread-like white rhizoids and flattened white scales.

STREAM BANKS

One of the habitats where mosses predominate is the damp slopes of stream banks. Because of the low light and the steepness, higher plants are less successful at competing here, and leaf litter does not accumulate to bury the bryophytes. Often definite zonation patterns exist that relate to the height of flooding during spring runoff or the rainy season.

Atrichum oerstedianum[IR,PR] (Ay′ trih kum　uhr steh dee ay′ nuhm) — **red crowned crane moss**, or **wavy catharinea**

While you are crawling along the stream bank on your hands and knees, peering at the world through a hand lens, you will think you have entered an Irish library on St. Patrick's Day. The large, lance-shaped (lanceolate) leaves of *Atrichum oerstedianum* have side by side stacks of cells running the length of the costa and looking like the pages of a book extending up from its binding. These neat little rows provide extra photosynthetic surface and most likely hold water in their capillary spaces.

This is one of the more common mosses used in Japanese moss gardens, where it is easily grown from leaf fragments. It is used in dish gardens to represent forests, and is often planted around bonsai. In nature, it is important in preventing bank erosion.

These plants form large masses on damp soil banks that get ample water, but when they get dry, their large leaves curl and twist so that only a small portion of the backs of the leaves are visible. When they receive sufficient moisture, these leaves will uncurl and begin absorbing both the rain and daylight.

Brachythecium rivulare[IR,PR] (Brack ee thee ' see uhm   riv yu lahr ' ee) — **rivulet cedar moss**

As the shallow water glides past the mud over bedrock near the edge of the stream, myriads of light green spikes penetrate the water surface and extend their blunt points toward the sun.   Closer examination reveals that these relatively large mosses are *Brachythecium rivulare*, sometimes curving gently at the tips, sometimes pointing straight upward.

These mosses can be rock builders, depositing calcium carbonate on their surfaces during photosynthesis, but in our area they are generally just bright green because too little calcium is available in the water.   They are usually abundant in seepage areas as well as stream borders.

Besides their large size, this moss can be distinguished by leaves that are concave, ovate, and possess a costa.   They most closely resemble *Platyhypnidium riparioides*, but if you examine the base of the leaf carefully, you will see that it extends down the stem on both sides, a feature that is generally not true of *Platyhypnidium*.

Dicranella heteromalla[PR] (Die crann ell ' lah   heh ter oh mal ' lah) — **green hair moss**

When you cross the stream, the bank is more than a meter (3 ') high and unstable.   Fine green hairs cover it and seem to be the only thing holding the soil in place.   This wispy moss is *Dicranella heteromalla*.

Dicranella heteromalla is among the most tolerant to sulfur dioxide of any moss known. It forms huge masses at the Sulfur Paint Pots, a geothermal area in western Canada. In the Upper Peninsula, it grows abundantly on soil banks in shade.

Because of its fine texture, it is used in dish gardens and moss gardens in Japan.

The leaves are very narrow and thin (green hair), curving like an eyelash. The capsules are usually abundant and their bright orange teeth are a striking contrast to the light brown capsules and yellow setae. The capsule itself is furrowed when dry and its opening is tilted.

Eurhynchium pulchellum[IR,PR] (Yu ring′ kee uhm   pull chell′ uhm) — **beautiful beak moss**

Growing near the *Dicranella heteromalla*, a moss with its lateral branches extending upward joins in holding the soil in place, safely above the high water mark of the stream. Apparently unable to tolerate the frequent covering by deciduous leaves, and perhaps intolerant of the dry pine woods, *Eurhynchium pulchellum* lives on soil banks, but seldom on flat ground, in deciduous forests.

The moss has a costa in its small, egg-shaped to nearly elliptical leaves. The leaf tips appear blunt because the pointed tip is usually bent over or twisted. It is easily confused with *Bryhnia*, but the leaves are more elliptical, not lance-shaped as in *Bryhnia*, and the moss usually has a clean, neat, shiny or bright appearance, whereas *Bryhnia* appears to be unkempt and dull.

Eurhynchium pulchellum

Bryhnia novae-angliae[PR] (Brinn′ ee ah noe vee ang′ lee ee) — **arrowhead moss**, or **common swallowtail moss**

This is an unkempt looking moss with twisted leaf tips. It generally grows on stream banks and near streams where it is constantly humid but not submersed.

Philonotis fontana[IR,PR] (Fill oh noh′ tis   fon tann′ uh) —
aquatic apple moss

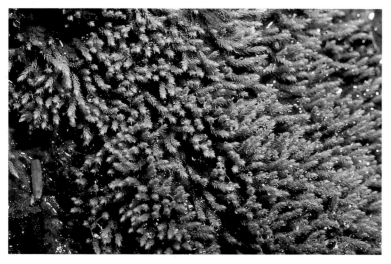

In seepy soil banks, around springs, along ditches, and across wet rocks, a whitish green moss with erect stems and a watery look forms expansive beds.   This intrusive moss is *Philonotis fontana*, looking its best while the water is abundant, and becoming rather dull when it gets drier later in the season.

Philonotis fontana branch.

Its abundance in the Himalayan highlands is partly responsible for its use there as chinking. The natives also use it to alleviate the pain of burns.

Its whitish green color when moist resembles that of *Pohlia wahlenbergii*, but its spherical capsules are quite different from the cylindrical ones of *Pohlia*. Under the microscope, it is possible to see bumps (papillae) on the ends of cells in *Philonotis*, but in the field, plants with no capsules can be difficult to identify. If the top of the plant has a flower-like structure with several branches protruding, it is *Philonotis*, most likely a male plant that has resumed growth after expelling sperm from the apex.

***Pohlia*[R,PR]** (Poh′ lee ah) — **spongy gourd moss**

Pohlia wahlenbergii in a waterfall at Copper Falls near Eagle Harbor.

The name gourd moss refers to the long neck of the capsule of *Pohlia*, distinguishing the genus from *Philonotis*, but not from its sister genus, *Bryum*. It differs from *Bryum* in having leaves with no border and in having elongate, rather linear cells. The leaves are generally long ovate to lanceolate (tear drop to long tear drop) with a distinct costa (rib).

Surrounding a gentle waterfall, more whitish green mosses form a lush mat as far as the splash can reach. Here one can find ***Pohlia wahlenbergii*[R,PR]** (wall en burr′ gee ee) — **pale-leaved thread moss**, or **paper lantern pohlia**, which is sometimes difficult to tell apart from *Philonotis fontana* in the field. Its whitish green leaves and reddish stems distinguish it from other *Pohlia* species in the field, but lose their striking contrast once collected and dried.

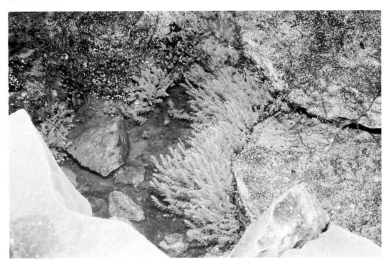

Pohlia wahlenbergii

Pohlia nutans[IR,PR] (Poh' lee ah   new' tans) — **nodding pohlia** — produces abundant elongate pear-shaped capsules, and in Mount Pleasant, Michigan, an observant bird-watcher found the stalks of the capsules lining a bird nest, glistening in the morning sun like threads of gold. The moss is very tolerant of sulfur dioxide, so it is not unusual to find it on sunny soil in polluted areas that are subject to little foot traffic.

Pohlia nutans (common on rotten stumps) has a golden sheen, *Pohlia cruda*[IR,PR] (on rocks or soil of damp banks or cliffs) has very shiny whitish, yellowish, or bluish-green leaves, and *Pohlia wahlenbergii* (typical of wet soil in splash areas) has reddish stems with whitish, watery looking leaves.

Plants with tiny plantlets or buds in the leaf axils are usually *Pohlia proligera*[PR] (proh lih   ger uh), and can be found on soil banks and in crevices of cliffs. The leaves are yellow-green and glossy, and the plants are small relative to *Pohlia wahlenbergii*.

Pohlia proligera with bulbils.

Conocephalum conicum[IR,PR] (Kahn oh seff′ uh lum   kahn′ ee kum) — **great scented liverwort**, or **snake liverwort**

Suddenly you draw back, fearing you have put your hand on a green snake, but realizing that it is a plant with a scaly skin look. There is a smell that is something between a cinnamon and a mushroom odor, causing you to pause and discover its source. You have encountered *Conocephalum*

conicum, a thalloid liverwort that is easily carried around on the bodies of turtles and other small animals, ultimately being planted along the stream banks.

In China, this liverwort is mixed with oils and used to treat cuts, burns, bites, boils, and eczema. It is known to inhibit the growth of micro-organisms, and thus might be of legitimate medicinal value. Unlike most bryophytes, which are avoided by herbivores, *Conocephalum* is eaten by slugs.

This is a large, thalloid liverwort. Its surface is divided into polygons, hence its scaly look. These each have a raised pore in the middle, resembling a miniature volcano. In the spring, a cone-shaped head (*cono* — cone, *cephalum* — head) on a tall stalk carries the female reproductive organs and ultimately the sporangia where the spores form. The bright green color of the thallus, its scaly appearance, and white rhizoids make this liverwort quite recognizable, but as its common name implies, it can also be recognized by its distinctive odor.

Pellia[R] (Pell′ lee yah) — **rivulet liverwort**

In small patches along the stream bank and in moist crevices of protruding rocks, a thin, dark green thalloid plant body extends long, white stalks to the heavens in great profusion. These delicate stalks, topped by a small black ball-shaped sporangium (capsule), mark the short reproductive period of the liverwort *Pellia*, so it is a rare treat to see them.

When the *Pellia* sporangium opens, it produces the 4-pointed star that is typical of liverworts. This genus differs from most of the other thallose liverworts discussed here in having no upright thalloid stalk to support its archegonia (female reproductive structure) or to keep its capsule in the air for a long period of time. Instead, the mature capsule is raised on a long, delicate, watery white stalk that soon disintegrates after the spores are dispersed.

Interestingly, in the right habitats members of this liverwort genus are able to contribute to rock-building by depositing carbonates on their own surfaces, yet continuing to grow upward, thus building up more and more rock.

Plagiochila porelloides[IR,PR] (Plah gee ah′ kill lah   pohr ell loy′ dees) — **common feather liverwort**

Just at the water line, forming a distinct zone above the *Fontinalis*, the large, leafy liverwort, *Plagiochila porelloides*, occupies the submersible zone along the stream.

Members of this leafy liverwort genus are used for chinking in the Himalayan Highlands. The plant contains substances that deter insects, and may some day be the source of pesticides.

The leaves lie in two rows and lack a costa, characters that are typical of leafy liverworts. These long elliptic, finely toothed leaves appear to be gathered at their insertions like a skirt into its waistband.

Preissia quadrata^{IR,PR} (Prize′ see uh   kwah dray′ tah) — **red marchantia**

A parade of tiny umbrellas marches across the ledge and down over the side of a calcareous rock. Although *Preissia quadrata* is not a *Marchantia*, as its common name implies (I suggest renaming it to red ribbon), it does have an umbrella-like stalked structure that carries the capsules. It is a thalloid liverwort that grows at the edge of streams, on rock ledges, and other places that are generally sunny for part of the day.

On the underside of the thallus, two rows of purplish-black scales, and often a purplish color to the underside of the thallus itself, distinguish this liverwort from *Marchantia*.

DUNG MOSSES

What better place to look for the dung mosses than on Isle Royale, where moose abound, and therefore their dung abounds. But don't despair that you must examine the foul smelling dung in order to enjoy this moss. The flies get to do that dirty work for you. By the time the moss is visible, the dung is dry and hard. This moss is in the only family of mosses known to be dispersed by flies. Flies are naturally attracted to dung, and so they get to deposit the spores there and begin the moss on its interesting and not quite understood life cycle.

Splachnum rubrum[R] (Splack' num rue' bruhm) — **umbrella moss, fairy parasols**, or **red collar moss**

The early botanists who found this dung moss were convinced that the iridescent purplish red parasol was a mushroom parasitic on the top of a moss. Now we know that it is not a mushroom, but is part of one of the most fascinating and unusual mosses in existence — *Splachnum rubrum*, which grows only on dung!

The capsules of *Splachnum rubrum* (left) and the leafy plants of *Splachnum* (right) on moose dung on Isle Royale.

The spores, looking like tiny pollen grains, apparently begin to grow on the moist dung. They form tiny green threads (protonemata), and as the threads grow, the dung dries and changes. Some months later, after the dung has turned into hard chunks about the size of charcoal briquettes, the moss forms its leafy shoots. By early July, the exquisite capsules (the parasols) begin to form. These tiny umbrellas are about 1 cm (0.5") across and have a brilliant metallic purple-red sheen that cannot help but catch your attention.

Actually, these tiny parasols have, just where you would look for the knob in the center of an umbrella, a yellowish cylinder that holds the new spores. But why should a fly visit the dung pile now, when the dung is old and odorless and covered with moss?

This moss has a special biochemical adaptation — it produces a musty sweet smell that attracts flies. Flies come to crawl around on the parasol, in the process getting some of the tiny spores on their bodies. Then, some of these flies visit patches of fresh dung of just the right age, and the tiny spores stick to the dung to begin the adventure anew.

Splachnum rubrum is easily characterized by the umbrella below the capsule, the brilliant capsule color, and thin, basal leaves with large cells. The leafy male plant (males never have capsules) resembles a green flower with a darker center and small teeth on the petal-like leaves. These delicate leaves soon shrivel beyond recognition when they get dry.

Splachnum ampullaceum[IR] (Splack' nuhm   am pew lay' cee uhm) — **common kettle moss**

A second member of this genus occurs in the bogs of Isle Royale National Park, and it may be seen exhibiting festive yellow and pink "kettles" or "lanterns" that are really its capsules with their swollen basal parts. Its leaves are quite distinctive because both male and female have large teeth that can be seen by the unaided eye.

Splachnum ampullaceum is more common in bogs and fens, where the humidity is constantly high, but *S. rubrum* may be found along the moose trails in lightly forested areas. *Splachnum rubrum* is unknown elsewhere in Michigan, and on Isle Royale it is restricted to growing on moose dung, so don't pick these rare beauties to grow yourself unless you think you are better at sniffing out the right dung for planting than the highly experienced flies!

Tetraplodon mnioides (Teh trap´ ploh don   nigh oy´ dees) — **brown tapering collar moss**

Like *Splachnum*, *Tetraplodon* is a dung moss, growing on owl pellets, carcasses, and dung of mammals. Members of this genus have been known from such strange substrata as the foot of an old stocking and the cloth of a half decayed hat. The most striking feature of *Tetraplodon mnioides* is the very crowded and conspicuous capsules. These flask-shaped capsules are reddish when young and become dark-colored at maturity. Male plants have star-shaped tops that resemble tiny green flowers. The leaves are long, ovate, lack teeth, and narrow abruptly into a long hair-like point.

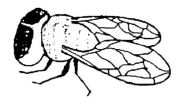

DECIDUOUS FORESTS

Since deciduous forests like maple and birch lose their leaves every autumn, you soon discover that the leaf litter covers everything, and mosses only appear on the slopes and protruding rocks, unable to survive under the dark cover of leaves and too small to penetrate their annual blanket. However, protected by the shade of summer, steep slopes can provide an interesting array of species.

Brachythecium[IR,PR] (Brack ee thee′ see uhm) — **green silk moss**

Brachythecium rutabulum is a common forest floor species that lacks plications in the leaves.

Invading lawns, rocks, and tree bases, or in the deciduous forest, *Brachythecium* species creep across the ground in numerous habitats. At least one species is credited with stabilizing sand dunes, partly because of its rapid horizontal growth and mat-forming habit. Others are important in Japanese moss gardens. They have been used for bedding, mattresses, cushions, and pillows, being preferred because they are insect repellent and resist rot. Their antibiotic properties contribute to their usefulness. In India, members of *Brachythecium* are used to wrap apples and plums.

Brachythecium species are among the few forest creeping mosses that have **plicate** leaves (folds that resemble accordion folds). Some leaves have teeth, but others do not.

There is a rib that extends at least half the length of the leaf, and the leaves are generally lance-shaped. As the name implies, the capsules (*thecium*) are short (*brachy*) and often purplish.

Dicranum scoparium[IR,PR] (Die kray' nuhm   skoh pair' ee uhm) — **broom moss**

Soft, round cushions dot the miniature landscape, looking like a forest of tall grasses in a strong wind due to sickle-shaped leaves that curve in one direction. Abundant in both the deciduous forest and Jack pine forest, *Dicranum scoparium* invites you to rest atop its cushiony growth.

It is among the most handsome of mosses in a Japanese garden, being especially important in the shade. In Europe, it is used in shop window displays, and its antibiotic properties enhance its use for diapers by American Indians.

This is one of the few mosses known to be eaten by mammals, being eaten by small mice and related rodents, but European rabbits dislike it, perhaps being more sensitive to the taste of the antibiotic compounds.

Dicranum scoparium can be distinguished from other *Dicranum* species by its large teeth, visible with a hand lens at the end of the leaf, and by ridges along the back (underside) of the costa. It is a highly variable species and can range in color from yellow-green in the sun to deep green in the shade. A curved capsule will distinguish it from some of its look-alikes.

Buxbaumia aphylla[PR] (Bucks baum′ ee uh   ay fill′ luh) —
bug on a stick, or **Aladdin's lamp**

As you examine the soil banks beside the paths on Isle
Royale, scanning for a sign of this moss, you will have made a
special find indeed if you succeed, because this moss has
been found on the island only once. This is an intriguing
scarcity, because it is common in the Upper Peninsula, and at
least some suitable habitats are on the island. It seems nearly
always to be associated with a tiny black leafy liverwort
(*Cephaloziella* — **small quill liverwort**), and I suspect that
there may be some type of cooperation between the species.

Buxbaumia is the delight of every bryologist. It lacks
leaves, and its thick stalk and large capsule have the intriguing
shape of Aladdin's magical lamp. The stalk and capsule
develop directly from the **protonema** (a threadlike structure
that develops directly from the spore). The capsule first
appears in late September, appearing waxy and green. It
orients its capsules so that the broad, flat surface faces south.
It survives winter in this state and completes its development
soon after the snow is gone, rapidly becoming brown. In this
state, it is very difficult to see because it blends well with the
ground. As spring turns to summer, the capsule usually splits
across the flat upper surface, and eventually the dry tissue on
that side breaks away, leaving the capsule looking as if it has
been feeding hungry forest creatures. This seems to be a
means of exit for the spores, as the operculum (lid) seems

often to remain attached well beyond the demise of a portion of the capsule.

Cephaloziella growing with *Buxbaumia aphylla.*

CalypogejaIR,PR (Cal ih poh jai′ uh) — **sack liverwort**

Members of this leafy liverwort genus are common on soil, if one crawls around and looks carefully. The tip of the leaf is somewhat obtuse or slightly pointed and often notched; this narrower tip gives the leaf the shape of a chicken egg attached obliquely to the stem. If one holds the stem at its base and lets the leaves hang down, the leaves overlap like the shingles of a roof when viewed from their upper surface (**incubous**), and this is the only local liverwort genus with unlobed leaves to overlap in this direction. Its common name is based on the fleshy, cylindric sac (marsupium) that hangs from beneath the branch and contains the developing capsule. In the field, *C. muelleriana* has a slightly bluish-green cast, caused by oil bodies in the cells, but this is lost when the plants dry and cannot be detected in herbarium specimens.

Diphyscium foliosum[PR] (Die fiss' kee uhm   fole ee oh' suhm) — **grain of wheat moss, powder gun moss**, or **nut moss**

While looking for *Buxbaumia*, you may find *Diphyscium foliosum*, because they often occur in the same soil bank where diffuse sun is available, but where they will not be trampled. This moss is common in the Keweenaw Peninsula in deciduous forests, but has not been found on Isle Royale.

Diphyscium foliosum has male and female plants that are very different. The female is most easily recognized when it has its strange stalkless capsules that resemble a grain surrounded by long hairs (leaves). The male plant, on the other hand, has unusual, long, strap-shaped leaves, and often fools bryologists who see it without the female.

Heterocladium dimorphum (Heh ter oh klay′ dee uhm die morf′ uhm) — **dimorphous feather moss**

Heterocladium dimorphum usually forms cushion-sized clumps on the forest floor. As its name (*dimorphum*) implies, this moss has two sizes of leaves, large ones on the stem and smaller ones on the branches. It is pinnate (arranged like the parts of a feather), with leaves that appear dull because of bumps on the cell surface (papillae).

Hylocomium pyrenaicum (Hy loh koh′ mee uhm pie rehn ay′ ee kuhm) — **mountain dragon-tail moss**, or **mountain pagoda moss**

But what is this? It looks like *Pleurozium,* but it feels different. This moss is a *Pleurozium* mimic, but it has small, leaflike **paraphyllia** along the stem and papillose leaves that give it a somewhat dull appearance. Unlike *Pleurozium*, it has a costa that extends up 2/3 of the leaf. It is not nearly so common as *Pleurozium* and grows on soil of mixed hardwood forests.

Hylocomium pyrenaicum

Leucobryum glaucum[IR,PR] (Lew koh bry′ uhm   glough′ kuhm) – **cushion moss**, or **powderpuff moss**

On the ridges and on sandy acid soil of lightly forested areas, especially among oak and pine trees, these whitish mosses can grow into large enough clumps to serve as an actual seat. Their leaves appear to be nearly white because they have several layers of cells, and many of these cells are colorless, thus hiding the color of the chlorophyll. The Japanese use it to cultivate rhododendron, and it has been reported to be used in a tailor's shop window for decoration. Of course it is used in dish gardens, where it can represent grassland, mountains, or landscape. It is used in shaded areas of moss gardens and as plantings around bonsai.

Plagiomnium ciliare[PR] (Plah gee oh nai′ uhm   sill lee air′ ee) — **many-fruited thread moss**

Suddenly, you stop to puzzle over a bed of green flowers on the bank at the side of the path. The leaves have a distinct border with single teeth consisting of 1 to 3 cells and extending nearly to the base of the leaf. But some plants have capsules drooping from the end of a stalk, and you must conclude that this "flower" really belongs to a moss. The male plants are separate from the female plants and form splash platforms that look like small green flowers at the shoot apex. Because of the obvious costa and ovate shape of the leaves, these male plants truly resemble miniaturized flowering plants. This moss is common on soil banks along paths.

Male plants of *Plagiomnium ciliare.*

72

Capsules of *Plagiomnium ciliare.*

Plagiomnium cuspidatumPR (Plah gee oh nai ' uhm   kuss pih day ' tuhm) — **woodsy mnium**

As you climb a streambank or trip over a log, you may see what looks like a tiny vascular plant with leaves coming from two sides of a horizontal stem.   This will most likely be a *Plagiomnium.* In *P. cuspidatum* and *P. drummondii*, an upright sexual branch ends in a flowerlike grouping of leaves surrounding the male reproductive organs. Extending horizontally are runners that have leaves on two sides, and each leaf has a very conspicuous costa, making the leaf resemble a tiny vascular plant leaf.

In China, a close relative of these mosses houses the Chinese gallnut aphid during the winter; in the summer the aphid is responsible for making the highly prized gallnut that is used for food.

Plagiothecium laetum[PR] (Plah gee oh thee′ see uhm   lee′ tuhm) — **glossy cotton moss**

Shiny and smooth as if it has been ironed by an elfin iron, *Plagiothecium laetum* nestles on soil between exposed tree roots or in tiny caves made by roots and soil banks. Its flattened leaves are folded near the edge and asymmetrical, with two short ribs at the base of each.

Thuidium delicatulum[IR,PR] (Thuh ih′ dee uhm   dell ih kat′ u luhm) — **delicate fern moss**

As you slosh through a path made soggy by spring rains, the delicate fronds of *Thuidium delicatulum* will indeed look like tiny ferns. The side branches are again divided, causing the delicate appearance. As a result of its fern-like appearance, this moss often appears in dish gardens, even in the U.S., and horticulturists use it for orchid culture. In the Himalayan Highlands it has been used for chinking.

The tiny, usually yellowish leaves appear dull because of bumps (papillae) on their cell surfaces, and the stems are covered with tiny branched threads (**paraphyllia**) among the leaves. A close relative, ***Thuidium recognitum***[IR,PR] (reh cog' nih tuhm), differs in having the tips of the main stem leaves bent back away from the stem. The paraphyllia differ, but this trait is visible only with the aid of a microscope.

TREE BASES

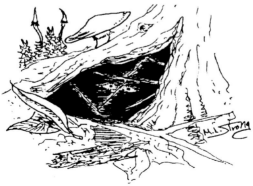

Since leaf litter covers even the most aggressive mosses on the forest floor, most species are relegated to the bases of trees and protruding rocks. However, nestled among the protruding tree roots there is a variety of mosses, some of which are found only in such a habitat. The nutrients from leaves and bark all funnel to the ground and can be readily absorbed by the mosses during a rainstorm. Moisture is greater than on any other part of the tree. The mosses themselves help to retain additional moisture, and most likely additional nutrients will splash up from the soil, while the mosses and liverworts need not compete with the higher vascular plants that cluster near the tree base in the soil.

Dicranum montanum[IR,PR] (Die kray′ nuhm mon tann′ uhm) — **mountain fork moss**

The elves must have grand parties here, because thousands of tiny corkscrews abound on the bark of the lower parts of the trees and logs and over the forest boulders. These corkscrews are leaves that belong to *Dicranum montanum*, and can serve as vegetative propagules to break off and start new plants.

These plants are dull, dark green and common. They are smaller than most of the other *Dicranum* species, and unlike most members of the genus, *D. montanum* has a straight capsule.

Dicranum flagellare[IR,PR] (Die kray′ nuhm flah jell lair′ ee) — **whip fork moss**

Close by, and easy to confuse with *Dicranum montanum*, is *Dicranum flagellare*. When it has its flagella, there is no problem. This moss sends up tiny, narrow shoots of closely appressed leaves that no other moss exhibits in our region. Like *D. montanum*, its capsules are straight.

Leskea gracilescens[PR] (Less' kee ah   grass ih less' sens)

Dull, blunt-tipped branches creep over the base of the tree, extending straight, cylindrical, rusty-colored capsules in all directions. The teeth of the capsules are white to light brown, making them conspicuous by contrast. The leaves are blunt-tipped, have a strong costa, and are rolled under (**revolute**) in the lower half, especially when dry. The species is somewhat rare in the Upper Great Lakes region, but it can be found at Pictured Rocks National Seashore.

Platygyrium repens[IR,PR] (Plat ee gire' ee uhm ree' pens) — **common flat brocade moss**

Bulbils of *Platygyrium repens*.

If you see an oily looking dark olive-green moss, yellow at the edges of the patch, creeping over the tree bases and on fallen trunks, it is most likely *Platygyrium repens.* This moss is most easily identified by its oily-shiny look and tiny **brood bodies** (short, deciduous branches) clustered at the tips of upright branches. These detachable short, budlike branches can be dispersed to start new plants.

TREE
TRUNKS

   While the tree base seems to be an ideal habitat, the trunk above is among the most severe habitats in a deciduous forest. It can be exposed to bright sunlight from the south, wind abrasion from the north, and rapid drying throughout. Nutrients are available only from dust and from leachates from the leaves and trunk above. This habitat is so severe that in most temperate forests no bryophytes grow on the branches, although lichens seem to be somewhat successful.

   One learns as a small child that it is possible to locate north because the mosses grow on the north side of the tree. In fact, this guidance is seldom reliable. Algae, and lichens that contain algae, are likely to be on the north side of the tree. Although bryophytes and lichens are typically more common on one side of the tree, factors such as wind direction, lean of the tree, and exposure contribute to determining which side. In Keweenaw County, I found that mosses in the first meter or so above ground were more common on the south side and quite sparse on the north side. I attribute this to the severe winds that desiccate and damage mosses on the north, whereas mosses on the south receive sunlight in winter, even through the snow, and are able to grow at the near freezing temperatures in the small space between the snow and tree trunk. Since the summer is generally too warm and dry for many species to grow, winter may be the most favorable season.

Metzgeria furcata (Metz gair′ ee ah   fuhr kay′ tah) — **forked liverwort, or ribbon moss**

Pressed against the bark in yellowish green sheets, this uncommon thallose liverwort will surely catch your eye. Its narrow (1-2 mm, 1/16″) thallus branches in pairs, forming Y's, with bristle-like hairs at the margins. It has a distinct midrib, which, along with its branching pattern, distinguish it from the liverwort *Riccardia*.

Neckera pennata[IR,PR]   (Neh′ ker ah   pen nay′ tah) — **feather flat moss**

Gorgeous shelves of pale green *Neckera pennata* extend from trees and rocks in the old growth forest. Each leaf is rippled like a lake on a breezy day. Its branches are distinctive in their flat appearance with elliptical, short-pointed leaves twisted to extend on only two sides of the stem. There are indications of a Stone-age human use of the moss *Neckera crispa* in a lakeshore settlement in West Germany. Even today species of *Neckera* provide plugs for seams and cracks of boats and canoes. Because of its beauty, one species has been used to make cords to decorate ladies' hats.

Neckera pennata

Orthotrichum[IR,PR] (Or thah′ trih kuhm) — **straight hair tree moss**

   Members of this genus defy their dry habitat, with their branches extended straight out in small tufts on tree trunks, or less commonly on rocks. The plants are dull and dark green or brownish with leaves crowded and appressed to the stem when dry, but not contorted. The name is derived from the straight hairs on the cap (**calyptra**). The capsule is usually partly hidden by the leaves because of the very short seta, but in a few species, the seta is elongate and the capsule extends above the leaves. There are 16 teeth in the capsule, often extending outward conspicuously. Recognition of the species is difficult, except for *Orthotrichum obtusifolium*, which has rounded, blunt-tipped leaves with flat margins. All the other species in this area have narrow, often pointed tips and **recurved** (rolled under) margins.

Porella platyphylla[IR,PR] (Pore ell' lah   plah tee fill' lah) — **false selaginella**

Elfin shelves, but this time on bark and boulders, and round like they have been made for individual plates . . . Branches with large, round leaves extend horizontally from the tree; leaves of this liverwort have smaller, oblong-ovate pointed leaf lobes on the underside, with a conspicuous underleaf that is twice as large as the underlobes and oblong-elliptical. Its nearest relative, ***Porella platyphylloidea***, has ventral leaf lobes as broad as the underleaves and almost circular. Like many of the leafy liverworts, this one produces a weak antibiotic.

Pylaisiella selwynii[PR] (Pie lay see ell' lah   sell winn' ee ee)
— **light green tree moss**

Look carefully at the tree trunks. You should be able to find a soft-looking yellowish green moss that looks neatly brushed upward as it creeps around the tree trunk. This would be *Pylaisiella selwynii*. It is easily distinguished by its small size (branches only a few mm long, 0.2") and light color. The only other moss in the Upper Peninsula with such distinctly upturned branches on tree trunks is *Leucodon brachypus* var. *andrewsianus*, and that moss is much larger (branches about 1 cm, 0.5" long), coarser, and has a dark green color.

Radula complanata[IR,PR] (Rad' yu lah   kahm plan ay' tah) – **flat-leaved liverwort**

In moist areas, one may find *Radula complanata*, which somewhat resembles *Porella*, but *Radula* lacks underleaves. It has small, round ventral lobes on the leaves and its rhizoids are produced on these lobes, not on the stems. It is a flat, yellow-green liverwort that is common on trunks of white cedar, but may also be found on the vertical rock walls of Jacobs Creek, both in moist areas.

Radula complanata

Ulota crispa^{IR,PR} (U low' tah   kriss' pah) – **curled leaf moss**, or **curled bristle moss**

Ulota crispa grows on tree bark and is especially sensitive to air pollution, so finding it in the Upper Peninsula and on Isle Royale attests to our wonderful clean air.

Ulota crispa (left) and its capsule with hairy calyptra (right, photo by Zen Iwatsuki).

84

The name *crispa* refers to the leaves that are curled and twisted when dry, but straight when wet. It is related to *Orthotrichum* and likewise has erect hairs on its calyptra. If you see tufts of dark green well above the base of the tree, with curled leaves, it is likely to be this moss. *Dicranum montanum*, which also has curled leaves, occurs near the base, but *Ulota* is adapted to the higher light intensity and drier conditions farther up the tree, and may even appear on some of the larger upper branches.

Frullania bolanderi[IR,PR] (Frew lay' nee ah   boh lann' der ai) **– millipede liverwort**

Clean air again favors this leafy liverwort that lives on tree bark. It is hardly noticeable on rough bark, but on smooth bark, especially light-colored bark, it is quite conspicuous. It is sufficiently common on bark to be a problem for some loggers. It has compounds that cause an irritation called wood cutters dermatitis, and sensitive persons will acquire an itchy rash. Even the wives of wood cutters can acquire the irritation by handling their husbands' clothing. The allergic reaction can be tested with a patch test similar to that used for other allergy-causing substances.

Frullania bolanderi (left) and its flagella-like small branches (right).

Members of the genus *Frullania* are usually dark reddish or brown, a color that most likely accounts for their ability to live high in the trees and in high elevations where they are exposed to intense radiation. The red pigment is able to absorb the UV light and prevent it from damaging the chlorophyll. Like all leafy liverworts, the leaves are in two rows, in this case with a third row of under leaves, and the plant grows horizontally. In *Frullania*, the leaves are round and have a smaller round, deeply concave lobe on the underside. In the Upper Peninsula, *Frullania bolanderi* is the most common species, and can be distinguished by small, flagella-like branches that are curved upward or away from the main plant body and that easily lose their leaves.

LOGS AND STUMPS

Just as tree bases favor many bryophytes, logs provide an oasis away from the scourge of leaf litter. Decaying logs furthermore can behave like a sponge and retain considerable amounts of water, thus providing a more favorable habitat than tree bases.

In this habitat, mosses provide additional protection against water loss, and particularly in coniferous forests, they nurture considerable seed germination. The fate of seedlings in this moss mat is less successful, partly because of competition, but also because the mosses themselves compete with the small seedlings for light. The system is further complicated by lack of soil for the roots, and if the log disintegrates beneath the young tree before its roots become well established in the soil, another deathtrap awaits.

Nevertheless, logs can be virtually covered with mosses and liverworts, some of which are almost exclusively in this habitat.

Aulacomnium androgynum (Awl oh kohm$'$ nee uhm   an droh jai$'$ nuhm) – **lover's moss**

Beware! A collection of miniature medieval clubs protrude from a patch of mosses — is there an army hiding inside? This pale green moss is distinctive because it always seems to have a narrow extension of the stem with round clusters of tiny brood bodies at the tip (and no brood bodies along this extension), resembling a medieval club. These brood bodies will readily grow into new plants, thus insuring the success of the population. The leaves have a costa, are coarsely toothed, smooth, and are usually not or only slightly contorted when dry. The stems grow in tufts on rotten logs, but are more common on soil or humus.

Bazzania trilobata[IR,PR] (Bah zay ' nee uh   try low bah ' tah) –
three-lobed bazzania

As you wander through a hemlock or boreal forest, old logs will often be covered with thick mats of this dark green leafy liverwort. As usual, it has its leaves in two rows, but they do not lie flat as in most leafy liverworts. Instead, the asymmetrical leaves drape gently down to each side of the stem, and each leaf has a broad tip with three small teeth. Unlike most leafy liverworts, the edge of the leaf closest to the tip of the branch overlaps the base of the next leaf (**incubous**),

88

so that if you look from the stem base toward the tip, they look like shingles on a roof, whereas in most other liverworts, the base overlaps the leaf behind it (**succubous**), and you must look from tip to base to see the shingle arrangement. Small branchlets arise from the underside of the stem.

Callicladium haldanianum[PR] (Kal lih klay' dee uhm   hal dane ee ay' nuhm) – **shiny moss**, or **common cedar moss**

By now you must be weary of all the hiking and new mosses to learn, so rest on a log and look for shiny golden green threads meandering gently over the logs and tree stumps. *Callicladium haldanianum* is a creeping moss that somewhat resembles *Brachythecium*, but its leaves are very flattened and smooth, giving them a shiny appearance. This moss has pointed leaves with no costa and no teeth on the margins. Its flattened appearance will alert you to its identity.

Dicranum viride[PR] (Die kray' nuhm   veer' ih dee) – **broken fork moss**

As you rest on the log, a bright green moss looking like miniature tufts of clipped grass brightens the landscape. This moss is distinguished by its ability to break off its leaf tips in a clean line across the leaf. These tips no doubt provide means of vegetative propagation for the moss. It is often found on decaying logs and tree trunks in the Keweenaw Peninsula and especially in the forests of Pictured Rocks National Lakeshore.

The only other local species with this unique ability to dismember the tips of its leaves is *Tortella fragilis*, which does not grow on logs or trees.

Dicranum viride

Hypnum imponens[IR,PR] (Hip′ nuhm   im poh′ nens) – **flat-tufted feather moss**

This curly moss looks like a miniature feather moss, especially *Ptilium crista-castrensis*, because of its neatly arranged branches. Its stem is orange and its leaves lack a costa, or have a short double one, thus distinguishing it from its other look-a-like, *Sanionia uncinata.* It is common on logs and tree bases, especially in the mixed hardwood forests of Pictured Rocks National Lakeshore.

Lepidozia reptans[IR,PR] (Lep ih doh′ zee ah   rep′ tans) – **creeping fingerleaf liverwort**

Although this liverwort is tiny, it is sufficiently distinctive to be recognizable. It has a very neat appearance because of the nearly perfect right angles formed by the branches and stem (remotely pinnate). Its tiny leaves are incubous, like those of *Bazzania*, but the plant is much smaller, and the leaves are deeply divided into 3-4 lobes, giving them a handlike appearance and the plant a filmy appearance. This liverwort tends to be a fresh, light green, and it occurs most often with other bryophytes on decaying wood, sandy soil, or rocks.

Lophocolea heterophylla[IR,PR] (Loh foh koh′ lee ah   heh ter ah′ fill lah) – **crested liverwort**

This very common liverwort is often seen among mosses and other liverworts, but the easiest place to find it is on the bare wood of rotting logs. Like its specific name implies, it has more than one leaf form (*heterophylla*), thus exhibiting oval, slightly lobed, and deeply lobed leaves all on the same plant. Its light green color and variable leaf habit make this liverwort easy to recognize. It is large enough to locate without a hand lens, but its variably lobed underleaves are only distinguishable with the help of a lens.

Lophocolea heterophylla

Oncophorus wahlenbergii[R] (Ahn kah' for uss wall enn burr' gee ee) – **bump moss**

The local creatures must surely be a party-loving lot, for here is another log with neat clumps of corkscrews. Small, curly mosses looking somewhat like *Dicranum montanum* and in some of the same habitats, these strange plants have a most unusual capsule. At the base of the capsule in the genus *Oncophorus* is a large bump resembling an Adam's apple! (or a tumor; oncology is the study of tumors) The capsule is furrowed, and in some species the teeth are bright orange, contrasting sharply with the light brown or beige capsule. Members of this genus in the the upper Great Lakes area are most likely to be found on logs in shaded, somewhat humid environments. Their sensitivity to air pollution limits their distribution.

Oncophorus (lower photo by Zen Iwatsuki)

Tetraphis pellucida[IR,PR] (Teh′ trah fiss   pell lew′ sih dah) –
four tooth moss

Dinner is served on an old rotten stump, and surely it is a
St. Patty's Day special of green biscuits in a green bowl.
These green biscuits are asexual reproductive structures
called **gemmae**, and this is the only moss to provide a leafy
cup (bowl) for its gemmae. Perched atop a leafy stem that
resembles a *Mnium* or a *Bryum*, these cups have tiny lens-

shaped bits of plant tissue that develop new plants once delivered to a suitable substrate. But when you can find the beautiful, slender capsules, the cups are either absent or only last year's dead and brown ones remain. Occasionally you may find only a naked group of gemmae when the leaves of the cup are no longer present or are just forming. As the generic name implies, this moss has but four teeth in its slender, erect capsule, making it unique among our mosses.

Although this moss is reputedly a lover of rotten wood, it is likewise common on sandstone cliff faces and even on soil banks high in organic matter, perhaps in places that once had decaying logs.

Tetraphis pellucida capsules showing four teeth.

EXPOSED SOIL OF UPROOTED TREES

This habitat hardly seems worthy of mention, but in ecosystems undisturbed by man, it is often the only habitat suitable for some species. It offers the advantage of more light exposure and cleared, mineral soil with no competition. Whereas species of higher vascular plants typical of open fields invade the disturbed soil of the ground, mosses invade the exposed soil nestled among the upturned roots. *Dicranella heteromalla*, already discussed as common on forest stream banks, is among these, but most notable, and not common elsewhere unless soil is disturbed, is *Leptobryum pyriforme.*

Leptobryum pyriforme[IR,PR] (Lep toh bry' uhm   pie rih forr'
mee) – **pear-shaped thread moss**

At the tip of each delicate stalk (seta) of this moss hangs a
pear-shaped green capsule, as its name implies. The threads
are the narrow leaves that look like fine green wires. These
leaves become longer near the top of the plant. This moss is a
pioneer that seems to abhor competition, thus being relegated
to such habitats as the upturned roots of trees, and flower pots
in greenhouses.

JACK PINE AND BOREAL FORESTS

Although the Jack pine (*Pinus banksiana*) forest is much
drier than the boreal forest of spruce and fir, the two share a
relatively acid forest floor with diffuse light and little broadleaf
litter. As a result, the species of mosses are similar, but the
abundances differ considerably.

Mature boreal forests usually have extensive moss cover,
whereas the Jack pine forest usually remains relatively barren.
Reduced growth rates of mosses may be due to the dryness
from the greater light penetration and often sandy soil of the
Jack pine forest, resulting in much less moss cover. Mosses,
however, can play an important role in the Jack pine forest. In
the southern United States in both recently burned and
unburned oak-pine forests, yellow pine (*Pinus echinata*)
seedlings are restricted to moss patches, apparently because

this is the only cover type that remains largely unburned. On the other hand, in the Jack pine forest, it seems that germination success of the Jack pine is high among mosses such as *Pleurozium schreberi*, but seedling success is low. This seems to be the result of competition for light, with the moss being the winner against the small, slow-growing seedling during the first year of the seedling's existence.

Abietinella abietina[IR] (Ah bee eh tin ell′ lah   ah bee eh tee′ nah) — **wiry fern moss**

A close relative of *Thuidium*, this moss has simple pinnate branching that gives it the appearance of a fern. Its leaves and branches are bronze-green or brown and dull due to bumps (papillae) that project from the cells. It is common in northern conifer forests, hence the name *abietina* (from the fir genus, *Abies*). It generally occurs on soil, helping to form overhanging soil banks at the margins of conifer forests. Thus, it is common along the Lake Superior shore in Keweenaw County.

Dicranum polysetum[IR,PR] (Die cray′ nuhm   pah lee see′ tuhm) — **wavy broom moss**

The generic name, *Dicranum*, literally means two heads, and the specific epithet means many setae, or stalks. The species, *Dicranum polysetum*, is unusual in possessing several setae that arise from the same apical portion of a single stem of this plant. Furthermore, as the common name implies, the leaves are very undulate, looking much like a sheet of paper that has been soaked by water and will never again lie flat. The leaves have a keel that is folded along the costa. The rhizoids of this species typically form a white or brown tomentum along a considerable portion of the stem. Compared to *Dicranum scoparium*, this species has leaves that are pale green or yellowish and usually shiny. It is the only one of our common *Dicrana* that has undulate leaves, except for the bog species *D. undulatum*.

This species is common in the same places as *Pleurozium schreberi*, and it seems to engage in a dynamic competition with that species, sometimes winning, sometimes losing, depending on the weather in several consecutive seasons. It is able to grow up through the *Pleurozium* plants because of its upright branching habit, but it cannot quickly extend horizontally. Once it has formed a thick and broad cushion or turf, it is not easily invaded, so winner or loser may be largely a matter of timing and chance.

Hylocomium splendens[IR,PR] (Hie loh coh′ mee uhm   splen′ dens) — **mountain fern moss, mountain feather moss**, or **stair step moss**

In damp fir forests, covering logs and forest floor like a feathery comforter, this moss invariably catches the eye of even the most casual observer.   Historically, *Hylocomium splendens* has enjoyed a variety of uses. Its large size made it useful for chinking in Alaska and fishermen's log cabins on Isle Royale, stopping leaks in log dams, moss gardening, and floral exhibits.   In India, it is used to cushion the weight of water vessels that women carry on their heads. Its sensitivity to SO_2 (sulfur dioxide) makes it useful as an air pollution monitor, but likewise makes it difficult to grow in moss gardens that are near areas of SO_2 emission.

Its common names remind us of its mountain or northern habit. It has the appearance of a fern, with primary and secondary branches that all lie in one plane. Each year, a new branch arises near the tip of the old one and branches upward, thus giving the appearance of a set of stairs with each step representing a year of growth.

The stems are reddish like those of *Pleurozium schreberi*, and under poor growth conditions it is possible to confuse the two. However, flattened branches will ordinarily distinguish this species.

Pleurozium schreberi[IR,PR] (Pluhr oh' zee uhm shray' ber ai) — **red-stemmed moss**, or **big red stem**

The most conspicuous and abundant moss in the Jack pine forest is the red-stemmed moss, which forms extensive mats of shiny yellow-green feather-like branches. It is one of the large mosses included in the group called **feather mosses**, and is also a very important ground cover in the boreal forest. Its stems are red, like those of *Hylocomium splendens*, but it differs from that moss by having a stem with branches and no secondary branches. The leaves are folded lengthwise like an accordion (**plicate**) and have a short, double costa (rib), most likely not visible even with a hand lens. The leaves are abruptly short-pointed and concave. This moss seems to play a competitive role in the forest by decreasing the success of young pine seedlings during their first year of growth, perhaps by blocking their light. It rapidly absorbs the potassium that washes off the trees and may serve as a filter for heavy metal pollutants, preventing them from reaching the roots of trees and herbs. Acid rain seems to benefit the moss since it is adapted to acidic habitats. Its role in maintaining humidity and retaining nutrients on the forest floor still needs to be explored.

Polytrichum[IR,PR] (Poh lih′ trih kuhm) — **hairy cap moss**

Have the elves stolen the locks of a Finnish maiden to cover their spears!? The genus *Polytrichum* gets its name because during early spring the capsule wears a stocking cap (**calyptra**) covering most of its head (**capsule**), and this cap is covered with hairs. Later, the moss loses its cap and exposes a small, angled capsule with a white or pale brown membrane at its tip. Hiding under this membrane are tiny spores that can come out around the edge of the membrane, falling to the ground to make new moss plants.

Hairy calyptras of *Polytrichum piliferum.*

Aside from *Sphagnum* this is the best known of the moss genera, and its members are among the most useful, with some species growing to more than 30 cm (1′) high in wet areas of bogs and low roadside areas. In Europe, *Polytrichum* species have been used to make carpet and curtain brooms and to weave into baskets. In the mountains of India, they are woven into welcome mats. In China, they are made into a tea to treat colds, and at one time the oil was extracted and used to beautify women's hair. The genus is common worldwide, and so it is not surprising to find several members growing in Japanese gardens and dish gardens. In fact, members of this genus are the most common ones grown commercially for private moss gardens.

The hairy cap moss can be recognized because of its long (1 cm, 0.5″) leaves and its capsules with four angles, like a long box with round caps on each end. If you have a hand lens, you can find tiny vertical rows of cells (**lamellae**) coming from the middle of the leaf and looking like pages of an open book. Since mosses have no roots, they must rely mostly on water from the air, and when the air is dry, they can easily lose water, without being able to get more from the ground as flowering plants can, so such structures as lamellae most likely help to conserve water while increasing the amount of photosynthetic tissue that has exposure to moist air for gas exchange.

Polytrichum juniperinum in spring with red splash cups beckoning raindrops to deliver their sperm.

Male plants of *Polytrichum* resemble flowers because they have tiny cups (splash cups) at their tips. These cups are made of specialized leaves and they house the sperm containers (antheridia). When it rains, the raindrops splash the sperm away from the male plant, as much as 1 meter away (3'). Some of these drops will land on the female plants and the sperm will swim to the archegonium to reach the egg and cause fertilization. If this happens, the stalk and capsule will begin to form on the female plant, and may take up to fifteen months to reach maturity.

Partly because of the bright male splash cups, members of this genus are quite attractive in gardens, making a good moss to place between stones of a path or in exposed sandy areas.

Stiff leaves, layers of vertical cells on the leaves, large leaf base that wraps around the stem (sheathing), and a 4-angled capsule will distinguish this genus from all others.

If you find the **juniper hairy cap moss (*Polytrichum juniperinum*[IR,PR]**, jew nih purr eye' nuhm), the lamellae may be hidden because the edge of the leaf is folded over them, helping to hold water in the moss when the air is dry. This folded over portion generally gives a bluish appearance to the leaves. The tip of the leaf is brown, distinguishing it from *P. piliferum*. *Polytrichum juniperinum* generally occurs on soil in open, dry areas.

Polytrichum piliferum[IR,PR] (Poh lih' trih kuhm   pie lih' fuhr uhm) — **awned hairy cap moss** — is best distinguished by its leaves, for, as the name implies, the tip of the leaf is extended as a white hair. This moss is smaller than our other *Polytrichum* species, and it has its leaf edges folded over the lamellae like those of *Polytrichum juniperinum*. *Polytrichum piliferum* is even more adapted for dry habitats than the other *Polytrichum* species, and it is common throughout the Michigan Upper Peninsula in pine barrens and on the sand dunes of Pictured Rocks National Lakeshore. It is important in stabilizing the Lake Superior sands.

In wet places such as bog hummocks and wet meadows, as well as in drier forests, you can find **great goldilocks**, *Polytrichum commune*[IR,PR] (kuhm myu' nee), although most bog members are **Polytrichum strictum** (strick' tuhm). *Polytrichum commune* does not have the edge of the leaf rolled over, so you can see sharp teeth along the edge, and the leaf is bright green. Its leaf base is broadly sheathing and yellow and the leaf tip is brown. Its capsule has a distinct ring just under the base of the square part of the capsule.

Polytrichum piliferum male splash cups (upper) and clump of *Polytrichum strictum* in a fen (lower).

Several other species without the edge of the leaf folded over are common, but recognition is difficult without a microscope.

104

Ptilidium ciliare^{IR,PR} (Till ih′ dee uhm   sill ee air′   ee) —
palmate liverwort

Reddish brown, sometimes bronze, sometimes crimson,
this is perhaps the most beautiful liverwort in the upland forest.
Its thin leaves are finely divided. It generally grows upright on
soil, often among mosses, and commonly spills over the soil
banks along the Lake Superior shore. The leaves are lobed
and divided about half way, with fine, threadlike divisions all
around the lobes. Its close relative, *Ptilidium pulcherrimum,* is
smaller, with leaves more finely and deeply divided, leaving
only about 1/3 of the leaf width undivided. It is the less
common species in the Michigan Upper Peninsula and grows
horizontally on bark at tree bases, particularly at Pictured
Rocks National Lakeshore.

Ptilium crista-castrensis^{IR,PR} (Till′ lee uhm   kriss′ tah - cass
trehn′ siss) — **ostrich plume moss**, or **knight's plume**

No better mimic of the plume of an ostrich exists in the
forest than this perfectly formed feather moss. Its pale green
leaves and branch tips curl under, just as the edges of an
ostrich plume. In good growing conditions, these branches
often sit upright in pure stands of perfect feather mimics.
Crista refers to a crest and *castrensis* to a soldier of the camp,
suggesting that the moss is suitable to be the plumed crest of
a knight. However, no such use has been documented.

One might confuse this species with *Sanionia uncinata*, but the plicate leaves have a short and double costa or none, whereas *Sanionia* has a long costa. *Ptilium crista-castrensis* often grows upright, whereas *Sanionia uncinata* does not, and the latter has a less perfectly shaped plume.

Rhytidiadelphus triquetris[IR,PR] (Rai tih′ dee ah dell′ fuss try kwee′ triss) — **shaggy moss**

And a shaggy moss it is indeed! The relatively large leaves seem to have no particular direction in mind, and they remind one of a child who just got up from a nap after sleeping in her best clothes. The moss usually has a light green, sometimes yellowish color, with slightly shiny leaves and orange red stems. The leaves at the ends of the branches look particularly rumpled because instead of forming the tight bud one expects of most plants, mosses included, the leaves at the end are wide-spreading as if the bud had been broken from their midst.

The leaves have sharp serrations, look rumpled with irregular plications (folds), and have two long, slender costae (ribs). Projecting cell ends on the backs of the leaves may sometimes give the moss a dull look and contribute to its "unkempt" appearance.

Rhytidiadelphus triquetris

Sanionia uncinata[IR,PR] (San ee oh′ nee ah   uhn sin ay′ tah)
— **hooked moss**, or **sickle moss**

As you lean back at the base of a tree or rest on a rock or rotting log in a coniferous forest, this moss may cushion your weary body. It somewhat resembles *Ptilium crista-castrensis*, although it is not nearly so tidy. The somewhat frondlike stems turn every which direction on the log, unlike the nearly unidirectional ones of *Ptilium*, and the side branches are mostly irregular in spacing and length. The leaves are plicate and have a costa that ends above the middle of the leaf. Each leaf is strongly curved and turned to one side, hence earning the moss the name of sickle or hooked moss. It is yellowish-green and somewhat shiny.

AFTER FIRE

A charred stump serves as a memorial to a forest of the past, ravaged by fire started by a lightning strike at the end of a dry summer.

Fires are natural parts of many ecosystems, and pine forests depend on them. Surface fires usually do little damage to the forest, and may even revitalize the forest growth. Young trees of invading deciduous species are eliminated and conifer cones open, releasing new seeds and replenishing the conifer forest. Fire is necessary to maintain Jack pine and boreal forests in areas where deciduous trees could otherwise invade.

If the canopy is opened sufficiently, and the competition is eliminated, a succession of bryophytes is one of the first events in re-establishment of the original ecosystem. It appears that these bryophytes are unable to tolerate any form of competition, and I suspect they may even be unable to tolerate some of the organic compounds in the soil. Within the first year, *Funaria hygrometrica* will almost invariably arrive, often accompanied by *Marchantia polymorpha*. These remain for only a few years and are joined and perhaps supplanted by *Polytrichum* and *Ceratodon purpureus*.

Funaria hygrometrica[IR,PR] (Few nair' ee uh   hi grow meh' trih ka) — **twisted-cord moss, Cinderella moss, predictor moss**, or **charcoal peddler**

The specific name literally means water-measurer, referring to the water-sensitive teeth and setae. But the more interesting names refer to the places where it measures water: beds of charcoal! This is one of the first plants to invade charred wood after a forest fire, appearing on charred wood in Yellowstone National Park in fewer than eight months after the fires of 1988.

In spite of this seemingly inhospitable habitat, *Funaria hygrometrica* has large, delicate-looking leaves on short stems. Since it grows much taller in culture media, I suspect it is deficient in magnesium, which is negligible in rainwater but present under the forest canopy, where it washes out of the canopy leaves. In its open habitat in nature, *Funaria* surely gets little of this essential growth element.

Funaria hygrometrica

The capsules are pear-shaped when fresh, but soon become asymmetrical and ribbed as they dry. Their long stalks (setae) are unusual in their gentle curvature, resulting in a capsule that hangs down from a gracefully nodding seta.

As you might guess from the long list of common names, it is both common and conspicuous throughout the world. The capsules of *Funaria* are always abundant, accounting for its rapid invasion of newly charred wood and making it a nuisance in greenhouse pots, for it will soon invade the lot of them and exert its rapid establishment as a competitive advantage over other mosses that are just beginning to get established. But give it a bit of larger vascular plant competition outdoors and *Funaria* soon pales in the reduced light, perhaps also refusing to grow in the presence of neighboring plant chemicals. Thus, *Funaria* is not truly a weed, but a pioneer that is there and ready when a new substrate is exposed, but disappearing again within two or three years.

Because it is easy to grow, *Funaria* is among the most studied of the mosses and has been an important source for understanding the action of plant hormones and developmental strategies.

Marchantia polymorpha[R] (Marr shahn' tee ah   pah lee morr' fah) — **common liverwort**, or **money liverwort**

Any student who has studied liverworts in a biology class knows about the thallus of *Marchantia*. Because this species is large and abundant, and grows readily in greenhouses, it is available for both microscopic and living study from any botanical supply house.

Some years ago, a group of college students, intrigued by the terminology of "thallus of *Marchantia*," decided to have some fun with a grand spoof. They called a famous New York newspaper and reported that **the thallus of *Marchantia*** was expected on the next flight into the airport. To make the story complete, one student and his attendants emerged from the plane, with the **thallus** duly decked out in royal red. Other students patriotically extended the red carpet to the stairs of the airplane and the reporters photographed and published the gala event. Soon after, when the reporters learned, to their embarrassment, of the spoof, they discovered to their dismay that there was no law against impersonating a thallus!

Marchantia polymorpha with gemmae cups.

The thallus of *Marchantia* resembles a table set for dinner. It is dotted with tiny bowls, called **splash cups**, that house bits of thallus. These plant bits, or **gemmae**, are capable of forming a new plant just like the parent. This asexual means of reproduction is responsible for its abundance in greenhouses, where water from the sprinkling system (usually a hose) easily spreads the gemmae about, providing rapid dispersal of the very hardy and effective gemmae. This is our only native species with splash cups, so it is easy to identify.

In the spring, the thallus extends two unusual umbrellas upward. One is a flat platform with 7-9 shallow lobes. On top of this platform are tiny openings that permit water to enter the male chambers (**antheridial chambers**). When it rains, the antheridia swell, finally exude their **sperm**, and the latter are splashed to the female plant (and lots of less useful places).

The female plant has a wonderful timing mechanism that permits its parts to be just the right place at the right time. When the male plant is dispersing its sperm in raindrops, the female umbrella has not yet risen to its lofty height, so the sperm readily splash downward to the awaiting **archegonia**. Here they swim down the neck of the archegonium and fertilize the egg. As the young zygote (fertilized egg) develops into an embryo, the female umbrella grows and extends upward. The lobes, holding the archegonia, elongate and form 8-9 fingers that roll under, thus placing the archegonia on the underside. Here the capsule develops, and when the spores are finally mature, the capsule hangs down, ready to drop its spores to the ground below.

Marchantia polymorpha male plant with splash platform.

Marchantia polymorpha female plant with spores.

The widespread occurrence of *Marchantia polymorpha* on newly burned or disturbed soil attests to the efficiency of dispersal in this liverwort.

SHADED ROCKS

It seems that the most interesting and rare of the mosses usually occur on damp or shaded rock faces (and most certainly on those that are just beyond our reach!). This perhaps is because these habitats are more rare, but it must surely also result from the adaptations necessary to survive such a severe environment. Although the lack of higher vascular plants is an advantage, this habitat is subject to the rigors of high or low pH without the advantage of organic buffers like those found in most soils. The vertical rocks support only mosses that attach with rhizoids or persistent protonemata (green threads that develop from spores), and usually preclude mosses that attain any large size. Small leaves, as well as small stature, help prevent water loss, and papillae seem to protect the moss, but the real role of papillae, if any, is still conjecture. I might suggest that papillae could reduce predation and filter the harmful UV light so that it cannot destroy the dormant chlorophyll during dry periods.

112

Anomodon attenuatus[IR,PR] (Ah nahm' oh don at ten yew ay' tus) — **common tree apron moss**

Making aprons at the bases of trees and hanging from calcareous rocks, *Anomodon attenuatus* derives its specific name (*attenuatus*) from its habit of making branches that are narrower at the tips. Like many large mosses, *Anomodon* has been used for chinking. Its most unique use, however, was as a smoke filter for tobacco.

Anomodon attenuatus is easily recognized by its drooping branches that get smaller toward the ends. The leaves are close together, smaller toward the end of the branch, blunt, and shaped like a human tongue, but with shoulders in the lower half. The leaves are papillose, giving them a dull look, and they have an easily discerned costa.

Bartramia pomiformis[IR,PR] (Barr tray' mee uh pahm ih form' iss) — **apple moss**

These are surely elfin apples on a stick, waiting to be drenched in caramel. *Bartramia* is named for John Bartram, a plant collector and explorer from Pennsylvania in the late 18th century. Because of the beauty of the soft, green cushions, *Bartramia pomiformis* has been popular in dish garden landscapes. Cushions up to 6 cm tall (2-3") cover rock ledges and steep slopes and cliffs, especially in canyons and other moist areas. The round capsules protrude from the cushion, looking like green apples when young; when they are older, they are slightly reddish brown, still resembling apples until they dry and become furrowed.

Bartramia pomiformis cushion (upper) and moist red capsules (lower)

The leaves are dull due to papillae on their cells, and the leaf shape is long and narrow. *Bartramia pomiformis* curls when dry, whereas its sister species, *Bartramia ithyphylla*, has leaves that remain rigidly erect. Both species occur in the Upper Peninsula, but *Bartramia ithyphylla* is rare.

Plagiopus oederiana[IR] (Playh´ gee oh puss ee deh ree ay´ nuh) — **slender apple moss**

The slender apple moss resembles its close relatives in *Bartramia*, and likewise grows on damp, shaded cliffs and crevices, preferring calcareous soil. Its leaves, when dry, are only somewhat curved, not strongly contorted, but it is best distinguished by the absence of papillae on the leaves, but unfortunately requiring the microscope to ascertain. Striations on the cell impart a somewhat dull appearance. Its dull, olive-green color and its gracefully curved leaves pointing at nearly right angles to the stem become helpful once you are familiar with the species.

Encalypta ciliata[IR,PR] (En cah lipp´ tah sih lee ay´ tah) — **extinguisher moss**

The closing moments of a cliffside ceremony are marked by exquisite, tiny candle snuffers covering green and brown candles on long pedestals. The candle snuffer is the calyptra (cap) that has a fringe at the base and completely encloses the capsule. The capsule itself is erect and appears like a candle on a stick because of its long, narrow shape.

Preferring a calcareous habitat, the moss is prepared for the desiccating conditions of its rocky home by possessing leaf cells with numerous papillae that dull the leaves and give them a waxy appearance. The leaf is wide and strap-shaped, with a blunt tip that has a small yellow point extending from it.

Upon drying the leaves become folded and contorted around the stem of the moss.

Encalypta ciliata growing with *Tortella fragilis* (upper) and with its candle-snuffer calyptra (lower).

Homalia trichomanoides[IR] (Hoh mal′ ee ah   trih koh mann oy′ dees) — **blunt fern feather moss**

Shiny golden green branches extend like awnings from the boulders. This relative of *Neckera* grows on rocks in the Keweenaw Peninsula and somewhat resembles a leafy

116

liverwort. Its leaves appear to occur from only two sides of the stem, are oval with rounded tips, and have an inconspicuous, single rib (costa). The shiny leaves have a smooth, oily appearance, looking like a plant that has been pressed or ironed.

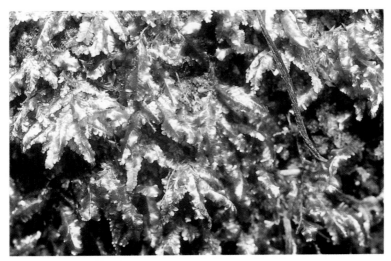

Homalia trichomanoides

Hymenostylium recurvirostre[PR] (Hie men oh sty′ lee uhm ree kuhr vih rahs′ tree) — **parrot moss**

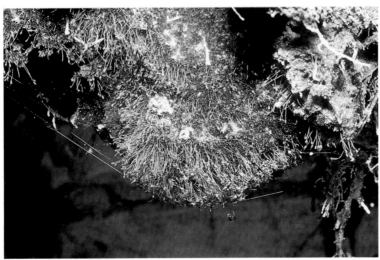

Abundant on moist limestone cliffs such as those at Pictured Rocks National Lakeshore, this species can be recognized by its peculiar **systylious** brown capsules, wherein the capsule lid (**operculum**) is perched atop a column (**columella**) that extends out of the center of the cylindrical capsule, where it remains for a long time after the capsule opens. The operculum has a long, curved beak, hence the name *recurvirostre*. The capsules have no teeth. The leaf cells have numerous papillae (projections), giving them a dull, waxy appearance. The base of the leaf is curled under on one side and the costa is strong, ending just below the sharp leaf tip. When dry, these leaves curl and are contorted.

Hymenostylium recurvirostre capsule, showing operculum perched atop the columella.

Paraleucobryum longifolium[PR] (Paah rah lew koh bry′ uhm lahn gih foe′ lee uhm) — **long-leaved moss**

From the view of a small elf, this looks like winds of hurricane force are bending every branch in sight. Everything seems to be curled in one direction, and the bluish cast of this plant reminds one of the silver backs of leaves upturned in the force of a pre-storm rustling of the wind.

This moss resembles a *Dicranum* except its leaves are several cell layers thick, slightly bluish or dull, and the capsules are straight. It is common on the boulders on Cliff Drive in the

Keweenaw and at Pictured Rocks on boulders that rest on the forest floor.

Paraleucobryum longifolium

Rhabdoweisia crispata[PR] (Rab doe why′ see ah   criss pay′ tah) — **small teeth coarse rock moss**

Such a small moss hardly deserves such a long (or absurd) common name, and in fact the scientific name is most likely easier to remember. The specific name, *crispata*, refers to its habit of curling its leaves when dry. This small moss forms tiny cushions on soil hanging from cliffs, helping these small sods to cling tenaciously to the steep slopes. It is most easily recognized by its abundant tiny capsules that are cylindrical, have 8 ribs, and usually have lost their teeth. When it grows on calcareous substrates, it can be confused with *Hymenostylium recurvirostre*, which is somewhat larger, forming larger cushions. The leaves of *Rhabdoweisia crispata* are not papillose, making them appear thin compared to the thicker, papillose leaves of *H. recurvirostre*.

Rhabdoweisia crispata when moist (upper) and dry (lower).

Rhodobryum ontariense[R] (Row doe bry' uhm   on tair ee enn' see) — **rose moss**

    Looking like a tiny green rose with its leaves crowded at the tip of a stem, this moss is sure to catch the attention of any who cross its path. It resembles a *Mnium* or *Bryum*, being related to the latter, and has a distinct costa and border. Its leafless stem grows from a horizontal **rhizome** (stem) and is covered by brown rhizoids.

Rhodobryum ontariense

In China, a member of this genus is one of the few mosses that has actually been tested to demonstrate that it has effective curative properties — it is used to treat cardiovascular disease and nervous prostration.

Saelania glaucescens[IR,PR] (See lay′ nee ah   glau kess′ sens) — **blue moss**

As you scramble up a soil bank along the Lake Superior shore, you may see a small bluish moss resembling a *Ceratodon* that has gotten covered by a fungus or lichen. It is *Saelania glaucescens*, looking bluish because of a white deposit on its surface that is apparently secreted by the moss itself. It has a simple upright structure and pale brown stalks (setae) with pale brown, erect, furrowed, cylindrical capsules.

Schistostega pennata (Schis toh stee′ gah   pen nay′ tah) — **luminous moss**, or **cave moss**

Who's there!?   As you peer into a cave, you may be startled by sparkling jewels on golden-green goblin threads shining back at you. These threads are the **protonemata** that develop from the spores of the luminous moss. Some of their cells are globose with their chloroplasts arranged on the side of the cell that is on the dark side of the cave so that the chloroplasts can absorb, and reflect, the maximum amount of light. This adaptation permits the plant to survive in places that have very limited light. This unusual moss has so intrigued the Japanese that there is an opera written about it and a monument celebrates its existence in Hokkaido.

Luminous protonema of *Schistostega pennata*.

The leafy plant has unusual leaves that lie in two rows and are not completely separated along the stem. This gives it a fernlike appearance, but for the tiny stalk and capsule at the end of the stem. These leafy plants are pale green and easily recognized growing in mini-caves, at rock bases, and among

122

tree roots, extending only 4-7 mm (0.2-0.3") tall. In Michigan's Upper Peninsula, this moss seems to prefer calcareous soil, occurring near Pictured Rocks National Lakeshore in a cave and abundantly at Tahquamenon Falls on soil in crevices of over-turned roots.

Leafy plants of *Schistostega pennata* with capsules.

Lophozia barbata[IR] (Low phow′ zee ah   barr bay′ tah) — **palmate maple liverwort**

Draped over rocks and soil, this common liverwort has the appearance of tiny fists strung along its stem. This fist appearance results because the dark green leaf is not flat, and its concave surface with a crimp between the lobes gives the appearance of knuckles. These leaves are attached obliquely. A spicy fragrance helps one to identify this liverwort.

Scapania nemorea[IR,PR] (Scah pay′ nee ah   nee morr′ ee ah) — **wood liverwort**

Hiding the sandstone face of the stream canyon wall, this common leafy liverwort derives its specific name of *nemorea* from its habit of living in groves of trees. Like *Porella*, it has underlobes, but in this case, the underlobes are larger than the upper lobes, resulting from a fold along the basal portion and making *Scapania* unique among our genera. The leaves are bordered by teeth, an uncommon character for liverworts, and often produce cinnamon-colored gemmae at their tips.

EXPOSED ROCKS AND SAND

Thank goodness for breezes that blow away the mosquitoes on the rocky ridges! Bryophytes and lichens are among the best-adapted plants for living in exposed, dry places. Both of these groups of organisms are able to remain dry for long periods, then regain their normal color and vitality in minutes after becoming wet. The plants that live here are often dark in color, perhaps using the dark pigment to protect their chlorophyll from damage by sunlight. The plants likewise often look dull because of tiny projections on their cells, called papillae, that seem to offer some protection in these exposed places, although no one is really sure how.

Andreaea[IR] (Ahndr ee′ ah) — **rock mosses**

As you scurry over a boulder, your hand becomes covered with tiny black fragments. You have damaged the brittle and fragile *Andreaea*. In our area, these are small, blackish mosses on rocks. Although this description is often used for *Grimmia* and *Schistidium*, most *Andreaea* species are even smaller and often, especially when dry, blacker. However, when moist, they may have a reddish-black or brown appearance.

Although *Andreaea* is a moss genus, it in some ways resembles a liverwort. The capsules split into four valves, like those of the leafy liverworts, not like the complex structure of a moss capsule. However, the leaves are in more than two rows, giving a moss appearance. The spores commonly divide once before being shed from the capsule, as do those of many leafy liverworts, and may help them to survive in their rocky habitats.

Two species of *Andreaea* occur in the Upper Peninsula. ***Andreaea rupestris*** (rew pess′ triss) lacks a costa and has fiddle-shaped leaves. ***Andreaea rothii*** (rawth′ ee ai) leaves are narrow and tapered and have a costa. Both species occur most commonly on acid rocks (conglomerates in our area) and are brittle and rigid.

Hedwigia ciliata[IR] (Hed wig′ ee ah   sih lee ay′ tah) — **white-tipped moss**

On dry rocks at a quarry at Pictured Rocks National Lakeshore, among the boulders on Cliff Drive in Keweenaw County, on rocky outcrops of Isle Royale, and elsewhere where rocks are exposed, this dry-looking moss appears to be frosted by an early season snow because of the white tips on its leaves. This frosty look is among its desirable traits that make it a good choice for growing on stones of moss gardens. When it is dry, the leaves are closely appressed to the stem, and the colorless cells at the tip of each leaf give the branches a whitened look, especially where the leaves are crowded at the stem tips. There is also a green variety in the UP that has no white tips.

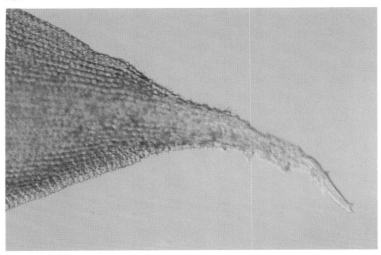

Hedwigia ciliata white-tipped leaf and papillae viewed with microscope.

Cellular projections (papillae) likewise contribute to the dry, whitened look, but when this moss is wet, it is dark green and its leaves spread quickly, ready to absorb the sun as soon as it returns, preparing for another siege of dryness between rainfalls. The capsules have almost no stalks and hide inconspicuously among specialized leaves on short branches.

Myurella sibirica[PR] (My uhr ell′ lah   sai beer′ ih kah) — **small mousetail moss**

Tiny, pale green mousetails protrude from small crevices of a rocky cliff! This moss usually occurs in small clumps with most stems being unbranched and pointing generally in one direction. The leaves are concave and oval, papillose at back and thus appearing dull, and result in a smooth, round appearance to the stem that resembles a mouse tail. The whitish appearance causes the genus somewhat to resemble *Bryum argenteum*, discussed later as a path moss.

Myurella julacea[IR,PR] (My uhr ell' lah jew lay' see uh) is similar, but the leaves are crowded and pressed closely to the stem, whereas in *Myurella sibirica*, one can see spaces between the leaves.

Racomitrium canescens[IR,PR] (Rah koh mih' tree uhm cann ess' sens) — **hoary fringed moss**, or **hoary wooly moss**

The whitened tips of leaves on this moss make it ideal for use in dish gardens to represent lakes, water, and snow. Its tremendous abundance in the Arctic has made it useful for chinking in Alaska and for lamp wicks in northeastern Labrador.

In Iceland, one can find acres of nothing but this species forming wonderfully soft cushions more than 30 cm (1') thick over the sharp lava rocks. The pale green color becomes gray when dry and looks much like the wool of a lamb, or one might imagine it as being covered with hoar frost because of its white leaf tips.

Our local populations generally have rather wide leaves and appear to be untidy because of numerous short, tuft-like branches.

Rhytidium rugosum (Rai tih′ dee uhm   rue goh′ suhm) — **wrinkle-leaved feather moss**

As you climb the rocky cliffs or stumble across piles of boulders in more or less exposed areas, you may find this large, sometimes regularly branched, golden-green moss with curved tips. Closer examination will reveal that the densely arranged leaves are deeply plicate and have a wavy, crumpled appearance, earning the moss the specific name of *rugosa*, meaning wrinkled.

Schistidium rivulare[IR] (Shiss tihd′ ee uhm   rih vu lah′ ree) — **common grimmia**

Exposed to the ravishes of Lake Superior winds and bright sun, this moss is a blackish rock dweller with its capsules hidden among the leaves and lacking a seta. Its teeth are bright salmon orange, and its costate leaves have small, squarish cells that contribute to its dark color because of their dark walls. This species is quite common where it can obtain some sunlight, and can be distinguished from the black *Andreaea* because *Schistidium rivulare* is not brittle.

Although *rivulare* means of a brook, this species has wide-ranging moisture tolerance. On rocks that are frequently wet,

"walk" on bare soil. He provided a control by walking on bare soil with no *Bryum argenteum*, then on more bare soil. *Bryum argenteum* grew abundantly on the soil that had been "walked on" by match books that had walked on *Bryum argenteum*, presumably because these were furnished with plant tips. Because of this ability to be dispersed by shoes, the moss is common in cemeteries along the walking areas. It actually benefits from being trampled.

Ceratodon purpureus[IR,PR] (Sur ah' toh don   pur pur' ee us) — **purple horn tooth moss**

Every spring, as you drive through the countryside, bands of green topped by reddish purple border the roads. Although purple readily describes the color of the mature seta and capsule, these soon change to a pale brown, leaving the moss with few distinguishing characters. I like the name one of my graduate students has given it, tricky moss, because it has fooled more bryologists than any other moss in the world. It has no single distinctive feature, except when its capsules are at their prime in early spring. The leaves are lance-shaped with their margins rolled under. When fresh and moist, their arrangement gives a star-shaped appearance when viewed from the top of the plant, and this may be slightly twisted like a pinwheel. When dry, they become more twisted and contorted, looking like they belong with *Barbula*. The capsule has a small swelling at the base, is slightly curved, furrowed, and has a beaked lid (operculum). Perhaps the most remarkable feature of the moss is its ability to grow almost anywhere. It grows on asphalt roofs, in cracks in the sidewalk, on rocky ledges, in open fields, and submerged in Antarctic lakes! (There is some debate as to whether the latter is the same species.)

Fissidens adianthoides[IR,PR] (Fiss′ ih dens   ah dee an toy′ dees) — **plume moss**, or **flat fork moss**

Layers of green feathers border the stream bank and path. Although the genus derives its name from the split teeth of the capsule, its most unusual features relate to the leaves. First of all, the leaves are flat, in two rows, and all lie in one plane. But even more unusual is the fact that each leaf has a pocket into which the next (newer) leaf base fits! No other moss has such a characteristic.

Fissidens adianthoides is our most common *Fissidens* and is distinguished by large and small teeth on the margin of the leaf. The plant seems to grow easily from fragments, thus accounting for its commonness along paths.

Polytrichastrum alpinum[IR,PR] (Poh lih trih kass′ truhm   al pie′ nuhm) — **alpine haircap moss**

Sausages on a stick appear to be awaiting an elfin hot dog roast. When no capsules are present, you will have difficulty distinguishing this from a *Polytrichum* in the field. Whereas *Polytrichum* has angular capsules, *Polytrichastrum alpinum* has capsules that are round in cross section, and the capsule resembles a broad, slightly curved cylinder, much like a cooked sausage or hot dog. The leaves have teeth on the edges, and the edges are not folded over as in *Polytrichum juniperinum* and *P. piliferum*. This species seems to prefer cool rocky cliffs, where it may form large clumps tenaciously holding mounds of soil. Its height and dark color make it useful in dish gardens to represent tall trees and dense forests.

DRY OR DISTURBED SOIL

You look sadly at the heap of soil where a new road has just been cut through the forest. But some bryophytes actually prefer disturbed areas and can be found nowhere else. These areas are temporarily devoid of competition, and the minute spores of the bryophytes are able to arrive more easily than the larger seeds of flowering plants. Thus opportunists thrive for several years, producing abundant spores, then disappear as higher vascular plants crowd them out and occupy the space and sunlight.

Barbula[PR] (Barr′ byu lah) — **beard moss**

The pale green leaves of this moss genus usually curl when dry but spread in fleshy-looking rosettes when moist. The most remarkable feature of the genus is its long, twisted teeth, like those of *Tortella*, that easily break off when dry. These are often bright orange or salmon-colored and quite striking.

Barbula plants are common in open areas, along paths, and in newly disturbed areas, keeping company with *Ceratodon purpureus* and *Bryum argenteum.*

Blasia pusilla[PR] (Blay′ zee ah   pew′ sill lah) — **common kettle liverwort**

A miniature leaf lettuce garden contrasts sharply with the red clay soil. This thallose liverwort grows on moist clay soil beside paths and in somewhat disturbed areas. Patches of star-shaped gemmae appear on the thalloid surface. This liverwort is intermediate between leafy and thallose, having leaves that are not completely separated from the axis. Under leaves are placed singly at the junction of the leaf lobe and the axis. The leaf lobes become filled with colonies of the Cyanobacterium *Nostoc*, which contributes nitrogen in a usable form to the liverwort. These *Nostoc* colonies appear as black points through the surface of the thallus, aiding in recognition of the species. Crum, in his *Mosses of the Great Lakes Forest,* does not report this species from lower Michigan, but it can be found on the red clays of the Keweenaw Peninsula and at Tahquamenon Falls.

Trematodon ambiguous (Tree mah´ toh don   am bih´ gyu us) — **crane moss**, or **long-necked moss**

This invader moss is rare simply because it will not remain once other plants invade. Therefore, it is relegated to a gypsy existence, constantly moving to a new sight of disturbance where it can live until its neighbors take over. Its long, usually slightly curved capsules are its most distinctive feature, sporting a long neck at the lower part of the capsule and shading from green to yellow to a salmon-orange as they age. The setae likewise are a bright orange and stand out in contrast to the surrounding green of leaves. The leaves are inconspicuous relative to the beautiful capsules and look like tiny green hairs because their base abruptly and quickly contracts into a long awn filled with the costa. Look for these on freshly disturbed sandy clay banks.

GLOSSARY

alkaline — having a high pH (above 7); calcareous

antheridium (pl. antheridia) — male reproductive organ; globose to broadly cylindric, stalked structure producing sperm

archegonium (pl. archegonia) — female reproductive organ; flask-shaped structure producing egg

basal — at the base

bog — waterlogged open type of vegetation poor in mineral nutrition because all water is derived from precipitation, rather than from the ground

boreal — northern; refers to northern forest of spruce (*Picea*) and fir (*Abies*)

border — margins differentiated from rest of leaf in shape, size, color, or thickness of cells

brood body — asexual reproductive body, usually made from modified plant organ such as leaf or branch

bryologist — person who studies mosses and liverworts

bryophyte — moss or liverwort; plant that has multicellular reproductive organs but lacks lignified vascular tissue

calcareous — having high calcium content

calyptra (pl. calyptrae) — membranous cap over young sporophyte, developed from tissue of archegonium, in true mosses ruptured near base, carried upward by elongation of seta, and continuing growth to form cap over capsule

capillary space — narrow channels that easily hold or absorb water

capitulum (pl. capitula) — rounded and compact grouping of branches forming head

capsule — spore case

concave — depressed in the middle

costa (pl. costae) — non-vascular nerve of leaf, sometimes double, sometimes single, sometimes absent

deciduous — shedding leaves annually, as the leaves of trees such as maple, birch, and poplar

decurrent — with margins extending down stem below leaf insertion as ridges or narrow wings

emergent — growing in the water, but extending above it

fen — open, boggy habitat receiving nutrients from groundwater, hence more mineral-rich than bog

flavonoid — compound that contributes color to plants

fluorescence — light produced while substance is being acted upon by radiant energy such as sunlight or ultraviolet radiation

gemma (pl. gemmae) — small, globose, elliptic, or cylindric body of few cells serving in vegetative reproduction

genus (pl. genera) — taxonomic grouping of related species

hummock — raised hump as found in bogs and fens

hyaline — colorless and transparent

incubous — leaves overlapping like shingles of roof when base of branch is at apex of roof

insertion — location where leaf base meets the stem

keeled — sharply folded along middle, like keel of boat

lamella (pl. lamellae) — in mosses, layer of cells that extends vertically from the middle of the leaf

lanceolate — lance-shaped, narrow and tapered from near base

liverwort — group of bryophytes including thallose and leafy members with top/bottom arrangement of plant parts; leafy members have leaves in two rows operculum (pl. opercula) — lid covering opening of moss capsule, usually falling at maturity to release spores

ovate — egg-shaped with base broader than apex

papilla (pl. papillae) — small protuberance, in mosses usually from cells

paraphyllium (pl. paraphyllia) — small, green filiform, lanceolate, or leaf-like, sometimes branched structures often produced on stems or branches of mosses

peristome — single or double circle of teeth inside mouth of capsule of most mosses

photosynthetic — any cell or tissue with chlorophyll; in *Sphagnum,* the green cells of the leaf

pinnate — arrangement whereby branches or leaves arise along two sides of an axis

plicate — folded in longitudinal pleats

propagule — reduced bud, branch, or leaf serving in vegetative reproduction

protonema (pl. protonemata) — green, branched filaments produced from germinating spores and giving rise to leafy or thallose plant

revolute — having leaf edges rolled under

rhizoid — simple or branched filaments, dead at maturity, anchoring plant and sometimes conducting water

rhizome — horizontal stem

seta (pl. setae) — stalk supporting moss capsule

species — taxonomic grouping of individuals that can all potentially interbreed and can readily be recognized as being related; members of a genus

sporangium (pl. sporangia) — spore-sac of capsule or, more loosely, entire capsule

spore — minute, mostly spherical, nearly always unicellular body produced in capsule

stained — colored with a dye to make some parts easier to see

succession — series of changes in plant communities leading to a community (climax) that remains stable in that geographic area

succubous — leaves overlapping like shingles of roof when tip of branch is at apex of roof

systylious — with operculum remaining attached to top of extruded columella after dehiscence

thallose — flat, not much differentiated, cordate or ribbon-like in body form

tomentum — dense woolly covering of rhizoids

underleaf — in liverworts, leaves on under side of stem; ventral leaves

undulate — wavy

vascular — having a system of conduction by which to move water throughout the plant

INDEX